The Outcomes and their Determinants from Community-Company Contracting over Forest Use in Post-Decentralization Indonesia

Charles Palmer

PETER LANG
Europäischer Verlag der Wissenschaften

Bibliographic Information published by Die Deutsche Bibliothek
Die Deutsche Bibliothek lists this publication in the Deutsche Nationalbibliografie; detailed bibliographic data is available in the internet at <http://dnb.ddb.de>.

Zugl.: Bonn, Univ., Diss., 2005

D 98
ISSN 0948-1338
ISBN 3-631-55196-7
US-ISBN 0-8204-9895-5

www.peterlang.de

Preface

In many developing countries worldwide, the decentralization of natural resource management has involved a full or partial transfer of rights over natural resources from the state to local communities and user groups. Furthermore, globalization trends have increased commercial pressure on natural resources in developing countries, leading to the presence of commercial companies in areas containing resources claimed by such groups. Often, commercial actors have been gaining access to natural resources and subsequently exploiting these through negotiating for community-held rights. Thus, the combined pressures of decentralization and globalization have implications for finding the proper trade-offs between the environment and the demands for development of some of the poorest groups in society. Moreover, rights over natural resources frequently remain weakly defined in formal law and poorly enforced by state institutions. This has led to a rise in conflicts and the phenomenon of local user groups and communities 'self-enforcing' their natural resource rights vis-à-vis commercial interests.

From 1997-98 onwards after the fall of President Suharto, Indonesia experienced one of the fastest and deepest processes of decentralization observed worldwide. These developments affected natural resource management, in particular the country's tropical forests. Indonesia has one of the world's largest forest stocks, covering tens of millions of hectares of its land area. Commercial logging has long been a major factor in the degradation of these forests. Post-decentralization, hundreds of forest-dependent communities were permitted to exercise their forest claims by negotiating logging agreements with logging firms in exchange for financial and in-kind developments, and despite weakly defined and enforced property rights.

This research, entitled 'The Outcomes and their Determinants from Community-Company Contracting over Forest Use in Post-Decentralization Indonesia' and carried out between May 2002 and May 2005, focuses on logging agreements between communities and logging firms in Indonesia. It adopts a new approach in attempting to quantitatively assess the impacts of decentralization on forest-dependent communities. Moreover, taking post-decentralization outcomes, this study analyses community-company interactions in a theoretically rigorous and quantitative manner. It is based on one of the largest field surveys of its kind undertaken in Indonesia. The quantitative and qualitative data from this survey forms the basis of the empirical analysis and the following discussion. The results from this study have clear implications for the role of local communities in the management of natural resources in developing countries where property rights are weak and commercial actors seeking to exploit natural resources are present. In particular, the results stress the importance of a community's ability to self-enforce its natural resource rights, which was found to be critical in improving both economic and environmental outcomes from the negotiated logging agreements observed in Indonesia. This ability could be improved, for example, through lowering community poverty rates and opportunity costs, enhancing a community's ability for collective action, and by making more explicit the links between community welfare and environmental damages from logging. These results are likely to be useful for policy makers interested in encouraging alternative and less-damaging land-uses for communities residing in forest areas in developing countries.

The book should be of particular interest to researchers, policy makers and students within the fields of international development, environmental and institutional policy and economics.

F.H J.v.B.

Table of Contents

Preface v
List of Figures xiii
List of Tables xvii
List of Boxes xix
Acknowledgements xxi
Executive Summary xxiii
List of Abbreviations xxvii
1 Introduction 1
2 Background and Institutional Setting 3
2.1 Indonesian Forest Policy and Related Institutional Changes 3
2.1.1 Centralization of State Forest Authority 3
2.1.2 Mechanized Logging: HPH Concessions 4
2.1.3 Decentralization 5
2.1.4 Small-scale Concessions: HPHH and IPPK 6
2.2 Logging and the Role of Communities 8
2.3 Background to East Kalimantan and Selected Research Areas 10
2.3.1 Geography and Governance 10
2.3.2 Population, People and Livelihoods 13
2.3.3 Economy and the Timber Trade 14
2.3.4 Timber Prices, Timber Markets and Taxes 15
3 Literature Review and Conceptual Framework 19
3.1 Comparison of the Impacts of Mechanised Logging on Communities Before and After Decentralization 19
3.1.1 The Political Economy of Decentralization 19
3.1.2 Conceptual Framework 21
3.1.2.1 Financial Impacts (Monetary and Non-monetary Benefits) 21
3.1.2.2 Social Impacts 22
3.1.2.2 i. Empowerment 22
3.1.2.2 ii. Social Costs from Rent-seeking and Conflict 24
3.1.2.3 Enforcement Costs 24
3.1.2.3 i. Enforcement vis-à-vis the Company 24

3.1.2.3 ii. Enforcement vis-à-vis Other Communities 25
3.1.2.4 Rent-seeking Impacts 26
3.1.2.5 Environmental Impacts 28
3.2 Determinants of Community Benefits from Post-decentralization Negotiations 31
3.2.1 Introduction to the Modelling Approach 31
3.2.2 Property Rights Formation 32
3.2.3 Community-company Bargaining Over a Logging Agreement 35
3.2.4 Theoretical Hypotheses 37
3.2.4.1 The Effects of Parameters on Individual Stages 37
3.2.4.2 The Total Effects of Parameters 39
4 Research Methods and Questionnaires 41
4.1 Development of Questionnaires 41
4.2 Community and Household Selection 46
4.3 Fieldwork 48
5 Community and Household Characteristics 51
5.1 Nearest Town and Market 52
5.2 Population, Homogeneity and Ethnicity 53
5.3 Neighbouring Communities and Forest Border Agreements 55
5.4 Employment and Livelihoods 56
5.5 Land Claims and Registration 58
5.6 Community Decision-making: Access to and Use of Community Forest 59
5.7 Organisations 60
5.8 Infrastructure before Decentralization 61
5.9 Third Party Assistance 62
6 Community Experiences before Decentralization 63
6.1 Monetary and Non-monetary Benefits from Logging before Decentralization 64
6.2 Protests and Conflicts against Logging Companies before Decentralization 65

6.3 Environmental Impacts from Logging Companies before Decentralization......66
7 Post-decentralization Negotiation Processes......69
7.1 Prior Knowledge of IPPK/HPHH Systems before Negotiations......69
7.2 The Role of Brokers and Subcontracting......69
7.3 First Steps: Scouting for Forest Areas and Initial Contacts......72
7.4 First Meetings, Negotiations and Participation......74
7.5 Formal Procedures for Application of IPPK/HPHH Permits......77
7.6 Decision-making and Agreements......78
7.7 Permits......79
8 Contractual Outcomes from IPPK/HPHH Negotiations......83
8.1 The Companies, Agreements and Market Structures......83
8.2 Length and Types of Agreement......84
8.3 Terms in the Agreement......85
8.4 Production-based Fees......89
8.5 Community Expectations and Renegotiations......90
9 Actual Outcomes from IPPK/HPHH Negotiations......93
9.1 Length of IPPK/HPHH Operations......93
9.2 Area Logged and Log Production......93
9.3 Monetary and Non-monetary Benefits Received......94
9.4 Firm Compliance and Community-company Conflict......96
9.5 Conflict with Outsider Groups......99
9.6 Environmental Impacts of IPPK/HPHH Operations......100
10 Distributional Aspects of Outcomes......103
10.1 The Distribution of Total Rents and the Role of Brokers...... Revisited......103
10.2 Community Fees......108
10.3 The Distribution of Community Fees......109
10.4 Intra-community Conflict from Rent Distribution......111
11 Comparison of Impacts of Mechanized Logging on Communities before and after Decentralization......115
11.1 Monetary and Non-monetary Benefits......115

11.2 Social Impacts 117
11.3 Enforcement costs 118
11.3.1 Community-company Conflicts..................118
11.3.2 Inter-community Conflicts 119
11.4 Rent-seeking Costs 121
11.4.1 'True' Self-enforcement', 'Opportunistic' or 'Useless' Blockading? 121
11.5 Environmental Costs 127
11.5.1 Changes in Environmental Impacts from .... Decentralization 127
11.5.2 Trade-offs between the Environment and Other Contractual Provisions 129
12 Determinants of Community Benefits from Post-decentralization Logging Negotiations 133
12.1 Operationalizing the Model 133
12.2 Econometric Analysis 136
12.2.1 Results for Hypothesis Set 1 137
12.2.2 Results for Hypothesis Set 2 139
12.2.3 Results for Hypothesis Set 3 141
12.2.4A Comparison of Results 144
13 Discussion 147
13.1 Negotiations and participation 147
13.2 Outcomes and equity 150
13.3 Notes on limitations and possible extensions 161
14. Conclusions and Outlook 165
List of References 173
Appendices 181
Appendix 1 Paired Samples Test for Differences in Proportions of Households receiving Financial and Non-monetary Benefits[1], from Logging Companies before and after Decentralization 181
Appendix 2 Household Perceptions of Forest Ownership before and ......

after Decentralization and the Results from Paired Samples Test for Differences in Perceptions 182
Appendix 3 Correlation Matrix for Factor Analysis using Discrete Variable for Firm Compliance 183
Appendix 4 Correlation Matrix for Factor Analysis using Continuous Variable for Firm Compliance........................183
Appendix 5 Paired Samples Test for Differences Local Environmental Impacts from Mechanised Logging before and after Decentralization. 184
Appendix 6 Descriptive Statistics on Parameter Proxies.......................185
Appendix 7 *Hypotheses 1 & 2*. Proxies Used, Expected Effects on Community Payoffs, and Two-sided Correlations between Community Payoffs and Explanatory Variables 186
Appendix 8 *Hypothesis 3*. Proxies Used, Expected Effects on Community Payoffs, and Two-sided Correlations between Community Payoffs and Explanatory Variables 188

List of Figures

Figure 2.1 East Kalimantan and Districts 12
Figure 3.1 Outcomes of the Conflict Game 34
Figure 5.1 Distance to Nearest Town and Marketplace for all Sampled Communities (km). 53
Figure 5.2 Numbers of family heads for sampled communities........ 54
Figure 5.3 Numbers of Neighbouring Communities for Each Sampled Community by District........ 56
Figure 5.4 Amount of Land Claim (ha) for all Sampled Communities........ 58
Figure 5.5 Community Decision-making over Access to and Use of Forest Resources 59
Figure 5.6 Proportion of Surveyed Communities containing Selected Infrastructure before Decentralization 61
Figure 5.7 Assistance from District Government to Communities in Malinau, Bulungan and Kutai Barat. 62
Figure 6.1 Total Length of HPH Experiences for Sampled Communities........ 63
Figure 6.2 Community and Household Perceptions of Environmental Impacts from HPH Operations........ 67
Figure 7.1 Community Participation in First Meetings and Further Discussions with Company Representatives. 76
Figure 7.2 Household Data on Decision-making with Respect to IPPK/HPHH Concessions 78
Figure 7.3 Location of Permit Owners 81
Figure 8.1 'Companies', Agreements and Market Structure........ 83
Figure 8.2 Length of Agreements by District........ 84
Figure 8.3 Proportions of Different Types of Agreements by District........ 85
Figure 8.4 Proportions of Agreements Containing Terms on Borders and Locations of Concessions, Plantation Development and Employment 85
Figure 8.5 Proportion of Agreements Containing Terms Relating to Various Logging Rules 87
Figure 8.6 Proportions of Agreements Containing Advance Payments of Varying Sizes 88

Figure 8.7 Proportions of Agreements Containing Obligations to Social Developments.....89
Figure 8.8 Promised Fees, Rp. per m3 of Production Net of Permit Costs......90
Figure 9.1 Length of Company Operations.....93
Figure 9.2 Variation in Community Financial and In-kind Benefits from Logging Agreements.....95
Figure 9.3 Problems with IPPK/HPHH Companies as a Proportion of all Communities within Each District.....97
Figure 9.4 Proportions of Community-company Disputes resulting in Actions against the Companies.....98
Figure 9.5 Proportions of Communities Having Problems with Outsider Groups since IPPK/HPHH Activities Commenced.....99
Figure 9.6 Causes of Problems between Communities as a Proportion of all Communities Sampled within Each District, in Duration of IPPK/HPHH Operations. 100
Figure 9.7 Community and Household Perceptions of Environmental Impacts from IPPK/HPHH Operations......101
Figure 10.1 Generalised Structure of Distribution of Negotiated Rent Flows: Malinau and Bulungan.....104
Figure 10.2 Generalised Structure of Distribution of Negotiated Rent Flows: Kutai Barat.....105
Figure 10.3 Identity of First Recipients of the IPPK/HPHH Fee Money.....106
Figure 10.4 Systems of IPPK/HPHH Fee Distribution.....110
Figure 10.5 Problems within Communities as Result of Fee Distribution (Community Data).....112
Figure 10.6 The Proportions of Communities indicating Intra-community Problems from Logging Fee Monies by Proportion of Sampled Households.....113
Figure 11.1 Perceptions of Forest Ownership Before and After Decentralization.....118
Figure 11.2 Intensity of Inter-community Problems From Logging Fee Monies by Proportion of Sampled Households per Community..120

Figure 11.3 Comparison of Household Participation in Community-firm Conflicts against the Proportions of Households reporting on Intra-community Conflicts as a Result of IPPK/HPHH Fee Money 123
Figure 11.4 Household Perceptions of Negative Impacts from Mechanised Logging Before and After Decentralization 128
Figure 13.1 'Fairness' and Intra-community Conflicts from Systems of Fee Distribution in Communities........ 158

List of Tables

Table 2.1 Price Comparison of Red Meranti Logs in Indonesia and Malaysia, 1999-2002 15

Table 2.2 Comparative Profitability of HPH, HPHH, IPPK Concessions and Illegal Logging (all in USD) 17

Table 3.1 Impacts of Mechanised Logging on Communities: A Framework for Analysis 23

Table 4.1 Structure of the Questionnaires 43

Table 5.1 Communities Sampled in East Kalimantan and Year of Current Settlement Foundation, September 2003-January 2004 51

Table 6.1 Monetary and Non-monetary Benefits of Mechanised Logging before Decentralization (for Communities reporting Positive Benefits) 64

Table 7.1 Generalized Steps Towards Agreement in the Three Surveyed Districts 71

Table 11.1 Financial and Non-monetary Benefits of Mechanised Logging before and after Decentralization 116

Table 11.2 Comparison of Contracts with and without Environmental Provisions 131

Table 12.1 Econometric Results on Probability of Community being able to enforce Property Rights (Hypothesis Set 1) 138

Table 12.2 Econometric Results on Determinants of Actual Fee Levels Paid (Hypothesis Set 2) 140

Table 12.3 Econometric Results for Total Effects Model (Not Including NEGHPH) 142

Table 12.4 Econometric Results for Total Effects Model (Including NEGHPH) 143

Table 13.1 Summary of Chapter 11 Results for Impacts from Decentralization 151

List of Boxes

Box 4.1 Motivation for Questions in the Community Survey 44
Box 4.2 Motivations for Questions in the Household Survey.............................45
Box 10.1 Fee Structures for Batoq Kelo and Mamahaq Besar (Sungai mas) in Kutai Barat (all Rp. per m^3) 107

Acknowledgements

I wish to acknowledge the following people, without whom this thesis would not have been possible:

Prof. Dr. Stefanie Engel at the ETH Zürich, who gave expert supervision, guidance, advice and motivation while at the Centre for Development Research, University of Bonn;

Dr. Lini Wollenberg, Dr. Krystof Obidzinski and other members of the Adaptive and Collaborative Management (ACM) team and Governance Programme at the Centre for International Forestry Research (CIFOR) for assistance, numerous discussions and practical insights;

Further insights and motivations were provided by Dr. David Kaimowitz at CIFOR, Professor Ramon López of the University of Maryland and Emeritus Professor David Pearce OBE of University College London;

Fieldwork assistance was provided by an excellent research team: Ms. Rini Kusumawati, Mr. Dodi Hernawan, Mr. Doni, Mr. Nopilus, Mr. Samsu and Mr. Yan Ngau. In the field, further assistance was provided by Dt. Iman Suramenggala and the staff of Yayasan Pionir, Yayasan BILKA and Yayasan SHK Kaltim;

All fieldwork and research was generously funded by the Robert Bosch Foundation;

This thesis was supervised and evaluated by Professor Dr. Klaus Frohberg and Professor Dr. Karin Holm-Müller, both of the University of Bonn;

I would like to thank my parents for constant support, and give a very special thanks to Ulrike 'Bu' Joras, for always being at the other end of the phone.

Finally, this thesis is dedicated to the memory of David Pearce, a great and inspirational economist who sadly and unexpectedly passed away during 2005.

Charles Palmer,

New York, January 2006

Executive Summary

In 1999, decentralization reforms resulted in changes to the management of natural resources in Indonesia. Governance in forests shifted from a centralized system to one controlled by local authorities. Newly empowered forest-dependent communities exerted property rights over customary (*adat*) forest, leading to direct negotiations with logging companies in exchange for financial and non-monetary benefits. This was despite community property rights being weakly defined in a legal sense. According to previous empirical research, communities appeared to have gained a greater proportion of timber rents compared to the pre-decentralization situation, although the benefits that flowed to communities from these agreements appeared to vary significantly.

This thesis sought to undertake two research objectives. First, to identify and compare the impacts of mechanized logging on communities from concessions given out before and after decentralization. Second, to identify and analyse the conditions that resulted in the variation in post-decentralization outcomes. A conceptual framework was developed for the first objective, and a game-theoretic model of community-firm interactions was adopted for the second. The conceptual framework on the impacts from decentralization considered five categories of impact: financial and non-monetary benefits, enforcement costs (conflict with firms and conflict with other communities), rent-seeking costs, social costs and benefits, and environmental impacts. The expected directions of effects were derived. Combining conflict and bargaining theory, the second model yields hypotheses on the factors determining the community's ability for self-enforcement as well as other determinants of community payoffs.

Given the widespread popularity of and participation in small-scale concessions in the immediate post-decentralization era, a relatively large number of communities were sampled in order to directly compare agreements. Community and household level questionnaires were developed to allow for such a comparison. Agreements were surveyed in three districts (Malinau, Bulungan and Kutai Barat) in a single province, East Kalimantan. Fieldwork was undertaken in 65 communities. A total of 65 community-level and 687 household interviews were conducted between September 2003 and January 2004.

The descriptive results show that the small-concessions system in Malinau and Bulungan were dominated by a small number of brokers, who were mainly from outside the communities. In contrast to Malinau and Bulungan, communities in Kutai Barat were more likely to deal directly with logging contractors, have more control over the process of negotiations and own permits. Community elites were responsible for much decision-making with respect to small-concessions and the distribution of rents. Active participation by community members in meetings about small-concessions did not, however, really translate into knowledge about the process of negotiations and awareness of what was contained within agreements.

The results on the impacts of decentralization showed that, as expected, significantly more households received financial and non-monetary benefits after decentralization compared to before. Communities only reported a slightly higher proportion of blockading activities against companies after decentralization compared with before decentralization, despite increased household participation. However, many of these activities may have been motivated ('opportunistic blockading'), misled ('useless blockading') or simply undermined by rent-seeking individuals in the community. A majority of households in over half of surveyed communities reported problems in the distribution of logging rents within the community. Anecdotal evidence collected in the course of the surveys suggested that distributional problems may have affected cohesion and trust among community members, and hence, the community's capacity for collective action. A majority of communities reported inter-community conflicts over forest borders and contradictory land claims as a result of post-decentralization concessions, although many of these conflicts originated in the past before decentralization. Nevertheless, similar to community-company conflict, they highlight the weakness and uncertainty of community property rights over the forest in the post-decentralization era. For environmental impacts, as expected, there were no significant differences between perceptions of negative impacts of logging on water quality, flooding or hunting, before and after decentralization. There were significantly less perceived negative impacts on the collection of forest products and farming, which was due to improved community access to the forest after decentralization. The results of a comparison of logging agreements with and without environmental provisions show no evidence of a trade-off between environmental and financial contractual provisions. Finally to measure community 'empowerment', a significantly higher proportion of households perceived community ownership over the forest after decentralization compared with before.

Regarding the determinants on actual levels of post-decentralization financial and non-monetary benefits, the theoretical hypotheses are supported through econometric analysis using survey data. In seeking to explain the factors underlying the observed variation in outcomes, the game-theoretic model suggests that where community rights remain weak and poorly enforced, the community's ability to self-enforce its rights over the forest is crucial for claiming a more significant share of logging profits.

The probability of a community effectively sharing in logging benefits by obtaining a level of actual payment above the minimal level (hypothesis 1) is found to be significantly increasing in the community's wealth (lowering discount rates) and household participation in existing village organizations (lowering self-enforcement costs). Also, the probability is increasing if the community's forest claim has been logged before and decreasing in the size of the concession area. For those communities able to self-enforce their property rights over the forest (hypothesis 2), the size of the payment received is strongly increasing in previous experiences of working for logging companies and decreasing in the distance to market. Considering the total effects of determinants on payoffs (hypothesis 3), the empirical results suggest that communities that valued the forest more, in particular those that derived a larger proportion of their income from the forest, were more likely to obtain higher payoffs. The same is true for wealthier communities, which have lower discount rates leading to greater ability to self-enforce rights and also greater bargaining power in negotiations. On the other hand, communities for whom self-enforcement was very costly—because they had high opportunity costs of time or low ability for collective action—were less able to claim an effective share in logging benefits. The results are consistent with collective action theory in that greater ethnic homogeneity and social capital appear to be associated with higher community payoffs. The impact of changes in the profitability of logging on actual payoffs is more complex. While higher profits increase the size of the cake to be shared, they also enhanced the firm's incentives to exploit the community through unilateral logging. Empirically, it was found that communities were able to obtain higher payoffs where concession areas were relatively small.

This is the first analysis of community-firm interactions to be based on quantitative data from a larger cross-section of communities, and the first quantitative analysis about the impacts of decentralization on forest-dependent communities in Indonesia. This study showed that while decentralization undoubtedly led to community empowerment, community forest rights need to be extended beyond simple 'use rights'. This is to reduce the incentive simply to log the forest as quickly as possible. Moreover, more clearly defined property rights over the forest also need to be made more secure. Also, this study highlights the importance of community self-enforcement in an uncertain legal environment. However, the model suggests that self-enforcement is a second-best solution to the protection of property rights. Given the inability of communities to monitor many of the negotiated provisions and high self-enforcement costs, a stronger and more impartial regime of government contract and hence, property rights enforcement may be more beneficial for communities.

List of Abbreviations

BIOMA *Biosfer-Manusia* (Peoples' Biosphere)
BPD *Badan Perwakilan Desa* (Representative Village Committee)
BPS *Badan Pusat Statistik* (Statistical Bureau)
CIFOR The Centre for International Forestry Research
CV Commanitaire Vennootschap (Limited Partnership)
FWI Forest Watch Indonesia
GDP Gross Domestic Product
GFW Global Forest Watch
Ha Hectares
HPH *Hak Pengushahaan Hutan* (Timber Extraction Licence)
HPHH *Hak Pemungutan Hasil Hutan* (License to Harvest Forest Products)
IDR Indonesian Rupiah
IDT *Impres Desa Tertinggal* (Presidential Instruction for Isolated Villages)
IPPK *Izin Pemungutan dan Pemanfaatan Kayu* (Licence to Extract and Use Timber)
KK *Kepala Keluarga* (Head of Household)
KKPKD *Kelompok Kerja Program Kehutanan Daerah* (Working Group Programme for Forestry)
KMO Kaiser-Meyer-Olkin measure of sampling adequacy
LPM *Lembaga Pengembangan Manusia* (Human Development Group)
MSA Measure of Sampling Adequacy
NGO Non Governmental Organisation
NTFP Non-Timber Forest Products
PKMT *Pembinaan Kesejahteraan Masyarakat Terasing* (Programme for the Improvement of the Welfare of Isolated Communities)
PMDH *Pembinaan Masyarakat Desa Hutan* (Development of Villages in Forest Areas).
PT. Limited Company
Rp. Indonesian Rupiah
RT *Rukun Tetangga* (Neighbourhood)
SASFMN Southeast Asia Sustainable Forest Management Network
SHK *Sistem Hutan Kerakyatan Kalimantan Timur* (Community Forest Management in East Kalimatan)
TNI *Tentara Nasional Indonesia* (National Army of Indonesia)
TPTI *Tebang Pilih Tanam Indonesia* (Indonesian Selective Cutting and Planting System)
WRI World Resources Institute
UNDP United Nations Development Programme
USD United States Dollars

1 Introduction

Since the fall of ex-President Suharto's centralized and autocratic government in 1997-98, the Republic of Indonesia has undergone rapid decentralization, resulting in changes to the institutions and processes governing the management of natural resources in the country (Barr and Resosudarmo, 2002). Governance over forests has shifted in a haphazard manner, from a centralized system of logging concessions and protected areas to one now informally controlled by over 300 district-level governments. Consequently, newly empowered forest-dependent communities have exerted property rights over customary (*adat*) forest, leading in many cases to communities negotiating directly with logging companies in exchange for access to financial and non-monetary benefits. For the first time ever, commercial timber harvesting operations have been paying fees and providing goods and services to communities in recognition of their presumed customary claims, although these are yet to be clearly defined in a formal legal sense (Wollenberg and Kartodihardjo, 2002).

In many countries, such as Indonesia, India, South Africa, Papua New Guinea, Ghana, and Mexico, decentralization and participatory approaches have led to situations in which companies that wanted to exploit forest products have had to negotiate with local communities with some rights over the forest (Mayers and Vermeulen, 2002; Bray et al., 2003). This is of particular interest in Indonesia, a country containing ten percent of the world's tropical forests, and possibly the third largest forest coverage in the world (FWI/GFW, 2002). Furthermore, forests cover 40 percent of the country's land area and support the livelihoods of an estimated 30 to 40 million people (FWI/GFW, 2002). Over the past five decades, however, forest cover has declined dramatically, from 162 to 98 million hectares (FWI/GFW, 2002). This deforestation was at least partly due to the rapid expansion of commercial, mechanised logging (World Bank, 2000; WRI, 2000).

Previous empirical research indicated that communities have benefited, at least financially and in the short-term, from the post-decentralization logging regime compared with the system before decentralization (e.g., Casson and Obidzinski, 2002). Before, the central government apparently only redistributed a relatively small proportion of timber rents from logging activities to the provinces. However, this research did not attempt to systematically assess the impacts of decentralization on Indonesian communities and similar to decentralization studies undertaken elsewhere (e.g., Larson and Ribot, 2004), has only offered a qualitative comparison of different case studies. Previous research, again typically undertaken on a case by case basis, shows a large variation in contractual outcomes and in actual benefits received by Indonesian communities from their negotiations with logging companies. For example, in the province of East Kalimantan case studies were undertaken by Barr et al. (2001), Casson (2001), Suremenggala et al. (2001), among others. Unfortunately, these qualitative studies are not directly comparable.

The research objectives of this thesis are as follows. The first is to identify and compare the impacts of mechanized logging on local communities from concessions given out before and after decentralization. The second is to identify and analyse the conditions that resulted in the variation in post-decentralization outcomes. A conceptual framework is developed for the first objective, and a more formal modeling approach is adopted for the second. Hypotheses on determining factors and the expected directions of effects are derived. Following the numerous case studies undertaken, it was decided that a relatively large number of communities would be sampled in order to allow for the direct comparison of contracts. This led to the development of community and household level questionnaires, combining qualitative and quantitative questions, to enable such a comparison. Moreover, it was decided to concentrate on agreements that had been made in three districts (Malinau, Bulungan and Kutai Barat) in a single province, East Kalimantan. The first two districts were selected on the basis of information that had been collected, in particular by Barr et al. (2001) and Suremenggala et al. (2001). Kutai Barat was selected due to reports of significantly higher promised fees made to communities compared with those reported in Malinau and Bulungan (see for example, Casson, 2001; and, Casson and Obidzinski, 2002).

The thesis is divided into four parts. The first part continues with background material (chapter 2), a literature review of the main concepts to be used (chapter 3) and the fieldwork (chapter 4). The conceptual framework and modelling in chapter 3 provide testable hypotheses and directions of effect applicable to the Indonesian context. Previous studies provided mainly anecdotal evidence on logging impacts, payoffs and conflicts. This study aims to build on these through the collection of comparative data from a much larger sized sample. The second part contains the descriptive results from this sample (chapters 5 to 10), while the third part presents the empirical analyses on the main research questions (chapters 11 and 12). Thus, given the complexity of the problem described in the second part, the third part attempts to abstract the main issues with a rigorous statistical treatment and econometric testing of the concepts and hypotheses presented in chapter 3. This kind of quantitative analysis using a sample of communities of this size has not been attempted before. The fourth and final part provides a discussion of the results (chapter 13), the study conclusions and an outlook to the future (chapter 14).

2 Background and Institutional Setting

In this chapter, Indonesian forest policy and related institutional changes including decentralization reforms are presented first. This is followed by the role of communities in logging. Next, background to the province of East Kalimantan is presented, including information on the study districts and timber markets.

2.1 Indonesian Forest Policy and Related Institutional Changes

Between 1957 and 1967, state forestry institutions in Indonesia were decentralized, resulting in jurisdiction over forests divided between the central and provincial governments (Ross, 2001). During this time, Indonesia's vast forests, spread out over up to 17,000 islands and covering perhaps up to 70-80 percent of the country's land area of over two million km^2, were yet to be commercially exploited on a large scale. This, plus the lack of state penetration and influence in large swathes of the country meant that its forests were *de facto* controlled by forest-dependent communities. Oral traditional and customary law (known collectively as *adat*) were the guidelines under which many of these communities managed the forest resource during this period (Ross, 2001).

2.1.1 Centralization of State Forest Authority

The first real move towards formal state control of Indonesia's forests was the enactment of the Basic Agrarian Law (1960), which formally categorised all customary (*adat*) forests as state forests. This law recognised customary property rights as long as these did not conflict with the 'national interest' (Ross, 2001). When Suharto and the New Order government came to power in 1966, Indonesia was undergoing economic crisis and in need of new sources of revenue (Resosudarmo, 2004). To this end, the adoption of a Basic Forestry Law (BFL) in 1967 centralized much of the state's authority over the forests, and swept away many of the legal and jurisdictional obstacles to large-scale commercial logging. All of Indonesia's forest land was placed under control of the central government in Jakarta, and customary land rights were subordinated to the forestry department's authority in order to promote commercial logging (Ross, 2001).

The forestry department issued timber concessions larger than 10,000 hectares (ha), while provincial and district governments retained the authority to grant smaller concessions. This division of authority allowed provincial and district officials to gain a fraction of the forest windfall. Perhaps inadvertently, it also allowed local smallholders, if not entire communities, to profit from the timber boom following 1967. This it did by creating concessions that were small enough to exploit with minimal capital, and could be acquired by local people. By 1970, over two million ha of these small concessions had been issued by local authorities (Ross, 2001).

2.1.2 Mechanized Logging: HPH Concessions

In 1970, the central government issued a new regulation[1] which set a new minimum concession size of 50,000 ha, thereby removing the ability of provincial officials to grant timber concessions. The rationale was that only large concessions could attract foreign investors. This regulation also stipulated that all logging be mechanized, putting non-mechanized operators out of business[2]. From this time, the central government was given the authority to grant exploitation rights (*Hak Pengushahaan Hutan* or HPH, a large-scale timber extraction concession licence) to private firms directly, without going through either the provincial governments or the forestry parastatal, Pehutani. HPH activities, through the payment of all manner of fees and royalties have been an important source of government income over the years. Moreover, HPH companies, unlike non-mechanized loggers, could be used to transfer rents to among others, the Indonesian military, in order to help consolidate the power of Suharto's regime (Brown, 1999).

In addition to the curtailing of legal small-scale logging in 1971, the government also adopted a regulation[3], which further weakened customary land rights. This stipulated that commercial loggers would have precedence over customary rights when the two were in conflict. Moreover, logging firms were given authority to regulate customary rights on their own concessions (Ross, 2001). By the late 1970s, timber concessions covered more than one quarter of the country. In these areas, the traditional rights of forest dwellers to use the land for farming ('swidden cultivation') were curtailed (GWF/FWI, 2002). Moreover, people were resettled away from forest areas, and had restrictions placed on the gathering of forest products or other uses of the forest (Lindayati, 2001). Thus, state forest policy from 1967 until the late 1990s was characterized by the belief that local land use practices were destructive and the resulting assumption

[1] Government Regulation (*Peraturan Pemerintah*) No. 20/1970.

[2] The forestry department stated that small-scale operators were difficult to regulate and failed to reforest logged-over areas. However, the fact that little capital investment was required meant that the timber from these concessions fetched lower prices than the larger, mechanised concessions (Ross, 2001). In 1971, the central government placed a complete ban on manual logging when the principle buying market, Japan, decided to stop buying any Indonesian logs produced by non-mechanized processes. This and a central government desire to shift concession allocations from the provinces to Jakarta led directly to the ban (Peluso, 1992).

[3] Government Regulation No. 21/1971.

for the need for a state policing role to limit local access and use (Natural Resources Institute, 1996). The Indonesian police and army (TNI or *Tentara Nasional Indonesia*) were paid to enforce all restrictions, protect HPH concessions, and ensure that noone else logged them (Casson and Obidzinski, 2002). In exchange, all HPH concession holders were mandated by central government to run rural development programmes for communities, known as *Pembinaan Masyarakat Desa Hutan* or PMDH ('development of villages in forest areas').

2.1.3 Decentralization

The commercial, mechanised exploitation of forest resources was a significant factor in fuelling Indonesia's rapid, though unsustainable economic development since the early 1970s (GFW/FWI, 2002; Friend, 2003). This lasted until the Asian economic crisis of the 1990s severely impacted the Indonesian economy in 1997, affecting the profitability of many large HPH logging companies and forcing some to abandon their concession areas altogether (Casson and Obidzinski, 2002). The economic crisis was accompanied by related social and environmental upheavals, leading to the rapid demise of the Suharto regime in 1998 (Forrester and May, 1999). At this time, there were widespread demands for political, administrative and economic reforms, commonly and collectively known as *reformasi*. Soon after, to show the government's intent to be more just and equitable with regard to forest resources, the Ministry of Forestry issued instructions in 1998 allowing communities to be more involved in forest exploitation (Casson and Obidzinski, 2002)[4]. These instructions allowed communities to benefit from low impact extraction activities, primarily of non-timber forest products (NTFP). However, as Casson and Obidzinski (2002) illustrate, many communities at that time were demanding more. They also wanted ownership rights to traditional forest areas as well as equal standing *vis-à-vis* HPH concessionaires regarding timber extraction.

In 1999, President Habibie's interim government passed new legislation on regional governance and on fiscal balance between the central government and the regional governments. The passage of two decentralization laws[5] led to the beginning of a fundamental political and administrative transformation of Indonesia (Resosudarmo, 2004). They gave greater financial and decision making

---

[4] Government Regulation No. 62/1998 and Decree from Minister of Forests (*SK MenHutBun*) No. 677/1998.

[5] Law No. 22/1999 on Regional Governance and Law No. 25/1999 on the Balance of Funds. See Dermawan and Resosudarmo (2002) for a full description and discussion of these laws and their implications.

powers to local government, particularly at the district and sub-district level[6]. Along with the legislation of the devolution of a wide range of public service functions, these laws also stipulated the transfer of natural resource management to the regions and increased the regions' share in natural resource revenues. For example, 80 percent of state income from resources (including forests) was to be retained by the regions (GFW/FWI, 2002). In the same year, a new Forestry Law (UU41) was introduced in 1999, replacing the 1967 law. This retained forest decision-making authority at the centre and established a category of forest known as customary forest (*hutan adat*), although the legal definition of community rights to forest areas remained very weak (Wollenberg and Kartodihardjo, 2001)[7]. However, there is widespread evidence that local people made decisions in the vacuum left by the loss of central government authority in 1997-98, suggesting that decentralization trends actually predated decentralization legislation, i.e. *de facto* decentralization occurred faster than *de jure* decentralization (Resosudarmo, 2004). For example, in Malinau in East Kalimantan, villagers had already begun to make claims to customary lands and negotiate directly with timber investors prior to the passing of the new Forestry Law allowing them to do so (Rhee, 2000).

2.1.4 Small-scale Concessions: HPHH and IPPK

The 1999 revised Forestry Law devolved elements of authority to manage forests from the central government to the provincial and district authorities, which included giving district heads the authority to issue licenses to harvest forest products (*Hak Pemungutan Hasil Hutan* or HPHH) and licences to extract and use timber (*Izin Pemungutan dan Pemanfaatan Kayu* or IPPK)[8]. These were to be valid for no longer than one year (unless renewed) and govern forest parcels no larger than 100 ha. The main beneficiaries of the permits were to be communities that lived in and utilized the forest in question and hence recognized many long-standing community land claims, typically for so-called customary forest (Suramenggala et al., 2001). Permits could in theory be assigned to individual land owners, village and government cooperatives, farmers' groups, community conservation groups and companies or other agencies (Barr et al., 2001). Casson and Obidzinski (2002) contend that HPHH were supposed to be allocated to

[6] Amid widespread perceptions of the central government's apparent inability to respond to the economic and political crisis, the decentralization laws were enacted relatively quickly in order to prevent 'national disintegration' (Van Zorge Report, 1999).

[7] Under the Law, the government is obliged to respect the rights of communities that have received its blessing as truly 'customary' but, as before, only as long as those rights do not conflict with the national interest (GFW/FWI, 2002).

[8] Minister of Forestry Decree No. 310 (1999).

those interested in extracting non-timber forest products using non-mechanized extraction techniques, while Palmer (2004) describes how these licenses were issued foremost for the small-scale harvesting of timber. IPPK licenses on the other hand were supposed to provide the means to open up or clear small forest areas for conversion to agricultural use or plantations. In reality, there appeared to be few differences between the two types of permit, as they were both used as a means of operationalising small-scale timber concessions, and can thus be directly compared with one another.

Following the increase in community forest claims in 1998-99, district leaders seized the opportunity to allocate large numbers of IPPK/HPHH licenses as they were seen as an excellent opportunity of raising district revenues (Casson and Obidzinski, 2002). In many districts in East Kalimantan, there was an explosion in demand for these permits, a demand that was at least in part driven by lower transaction costs for firms, compared with the centralised, HPH permit application process (see for example, Barr et al., 2001). By August 2000, the district head of Kutai Barat had allocated 223 HPHH permits covering 22,300 ha (Casson, 2001), a number that rose to 622 by December 2000 (KKPKD, 2001). Between August 1999 and August 2000, IPPK permits covering just under 10,000 ha were issued in Bulungan (*Bagian Perekonomian*, 2000), rising rapidly to 59,200 ha by the end of 2000 (Suramenggala et al., 2001). These 592 permits were spread among 42 companies and 30 communities. By the end of 2001, this figure had increased further to 618 permits and 61,800 ha. The district head in Malinau began issuing IPPKs in April 2000, with permits issued covering a total area of 56,000 ha by February 2001 (Barr et al., 2001). Throughout 2001 and 2002 small concessions continued to multiply, reaching for example as much as 80,000 ha in Kutai Barat (Casson and Obidzinski 2002). Resosudarmo (2004) notes that the Ministry of Forestry estimated the total area of Indonesian forest allocated for small-scale concessions by district governments since the system was established, to be in the order of two million hectares in January 2003[9].

Many IPPK/HPHH concessions were placed by district governments within the boundaries of HPH concessions, whether logged or not (e.g. Barr et al., 2001; Casson, 2001)[10]. This caused many of the larger concessionaires to complain and lobby the central and district governments for action against the newcomers, although some HPH operators were actively involved in IPPK/HPHH

---

[9] However, the decoupling of local and central government bureaucracies and the collapse in data gathering throughout the chain of governance implies that this estimate should be treated with caution (see Obidzinski and Palmer, 2002).

[10] HPH concessionaires were supposed to follow the Indonesian Selective Cutting and Planting System (*Sistem Tebang Pilih Tanam Indonesia*, or TPTI), with rotation times spanning decades. This usually resulted in some commercial timber stocks remaining in concession areas.

deals as well[11]. In response to this 'threat' to HPH concession areas and a realization that it had lost control over the allocation of IPPK/HPHH permits and was losing large amounts of potential revenue from the districts, the Ministry of Forestry suspended the authority of districts to issue IPPK/HPHH permits in 2000[12]. The official rationale was that IPPK/HPHH concessions were encouraging 'illegal logging'. However, many districts disagreed and justified their position based on differing interpretations of conflicting laws and regulations. According to district governments, small-scale concessions were 'legal' under the decentralization laws. Meanwhile, the Ministry of Forestry deemed these concessions 'illegal', basing its position and actions on the revised, forestry law (Larson, 2004)[13]. Regardless of the legal status of IPPK/HPHH, many districts continued to ignore central government decrees, issuing permits at least until 2003 when the research for this study was undertaken[14]. In 2003, a new concessions system, combining the IPPK/HPHH and HPH, was established to replace the IPPK/HPHH system (see chapter 14).

2.2 Logging and the Role of Communities

In the late 1960s in East Kalimantan, the growing international market for tropical hardwood timber, led to local operators organizing local chainsaw workers (including indigenous people) to log trees by hand (SASFMN, 1993). From 1968 to 1971, many smallholders employed a non-mechanized logging technique called *banjir kap* ('cutting through the flood') (Ruzicka, 1979)[15]. During

[11] These tended to be HPH concessionaires whose licenses had expired or who could not get their old concessions renewed (GFW/FWI, 2002). Others were looking to expand their concession areas but without the higher tax burden and bureaucracy associated with the HPH system (Casson, 2001). This not only improved the profitability of existing concessions but involvement with the IPPK/HPHH system also made them popular with local people as well.

[12] Minister of Forestry Decree No. 84/Kpts-II/2000.

[13] The implementation of Indonesia's decentralization laws have been marred by bad preparation and planning, leading to inconsistencies and ambiguities within and among different laws and regulations. Indeed, since 1998, even the basic division of authority and responsibility over forests among central, provincial, and district governments has been unclear and has been contested in many parts of Indonesia, including East Kalimantan. See Dermawan and Resosudarmo (2002) and Resosudarmo (2004).

[14] In addition to being popular among local people and raising district revenues, direct self-interest was also a factor in prolonging the IPPK/HPHH permit systems. For example in Kutai Barat, the district head apparently used some of the income generated to fund his election campaign (Casson, 2001).

[15] Since no roads or heavy equipment (trucks or bulldozers) were available, channels were dug to float the logs down to the river or to tributaries large enough to carry them during the wet season. Water buffalo were often used to haul the logs to the channels (Ross, 2001).

this time, non-mechanized loggers accounted for 62 percent of Indonesia's timber production and local people found non-mechanized logging more lucrative than working for large concessions (Ross, 2001)[16].

After the outlawing of small scale concessions in the early 1970s and their gradual marginalization by the state, local communities developed a strong mistrust of the government (GFW/FWI, 2002). Consequently, state forestry laws had little legitimacy in the eyes of local people. Given the kinds of institutionalized relationships found at the district level, the power of the Indonesian state to implement laws, especially in remote districts, was limited (McCarthy, 2000). This suggests that, even before decentralization, the management of local resources was already localized to some extent, with district elites operating with a large degree of independence from central government supervision. According to Casson and Obidzinski (2002) and Ross (2001), some community members involved in covert agreements with local logging entrepreneurs and HPH concessionaires, resulting in an informal timber economy. This tended to occur in areas where non-mechanised logging was already established prior to Suharto's centralization policies. After 1971, timber from these operations mainly served local demand. Thus, local leaders allowed loggers to cut community forests and assisted concessionaires in logging outside their boundaries or within protected areas in exchange for cash. McCarthy (2000) asserts that small-scale, non-mechanized logging by local communities occurred on a 'wide scale' prior to 1997 despite attempts to outlaw it. Dudley (2001) on the other hand states that what he called 'illegal logging' by communities remained on a minor scale because of the willingness of the police and military to enforce the law and, by extension, the interests of the large-scale logging companies. Regardless of the scale, the deliberate marginalization of local logging networks through the Suharto period probably never really completely eliminated them altogether.

Covert agreements between local logging entrepreneurs, communities and sometimes, HPH concessionaires, continued into the post-decentralization era with financial benefits flowing to some community members (Casson and Obidzinski, 2002). These agreements tended to focus on the non-mechanized felling of rare, high-value tree species such as Ironwood (see Limberg, 2004), sometimes only found deep in inaccessible forest. Moreover, this practice continued to be classified as 'illegal logging'. By contrast, those local entrepreneurs who were able to operationalize IPPK/HPHH concessions subsequently had their businesses 'legitimised' (see Obidzinski et al., 2001; Casson and Obidzinski, 2002).

[16] In the late 1960s, there was still a relatively high abundance of unexploited, high value, natural forest located in easily accessible locations near waterways and markets, which minimised the costs of non-mechanised operations.

The IPPK/HPHH system enabled many local timber brokers or 'entrepreneurs', already embedded in local illegal logging networks to expand and mechanize their operations, significantly increasing the levels of their rents from logging (Limberg, 2004). However, in order to expand operations and take advantage of the small-scale permit regime, local brokers instigated collaborations with capital owners and timber buyers forming working or 'limited partnerships' (known as *Commanitaire Vennootschap* or 'CV'). The new mechanised and legitimised operations were able to harvest far higher production volumes and hence comparably higher absolute levels of fees than manual logging. So while communities derived up to two-thirds *less* in IPPK/HPHH fees per cubic metre compared with manual logging, they received more money overall (Obidzinski, 2004). In addition to this financial incentive to participate in the small-scale permit regime, Casson and Obidzinski (2002: 2142) describe how communities recognised that this was an opportunity for them to 'legitimately' benefit from their forest claim and hence, obtain a legal recognition of their claims. Nevertheless, lacking the required capital and expertise, communities still had to collaborate with brokers and outside logging contractors in order to participate (Resosudarmo, 2004). Casson and Obidzinski (2002) observed that the upshot of community participation in IPPK/HPHH deals was increased vigilance and awareness of territorial borders, which apparently led to a decline in opportunities for freelance or non-mechanised felling of timber as well.

2.3 Background to East Kalimantan and Selected Research Areas

2.3.1 Geography and Governance

The province of East Kalimantan is located in the Indonesian part of the island of Borneo, to the south-east of the Malaysian state of Sabah. Of 33 provinces in Indonesia (as of September 2004), East Kalimantan with an area of 211,400 km^2, is the second largest province after West Papua (BPS, 2000)[17]. Prior to decentralization and regional autonomy, East Kalimantan was divided into four administrative districts. In the mid-1990s, the provincial government of East Kalimantan initiated the process of dividing up the districts of Bulungan and Kutai. The official rationale for this partition was to improve the capacity of the district governments to administer the areas within their jurisdictions and to provide essential services to a broader spectrum of their respective populations (Barr et al., 2001). In 2002, the province comprised eight administrative districts (see figure 2.1).

[17] In the near future, East Kalimantan may become the largest province in Indonesia if the plan to divide Papua into three separate provinces is implemented.

Prior to their partition in 1999, the original districts of Bulungan and Kutai were the largest districts in East Kalimantan (Bappeda Tk II Bulungan, 1998). The division of the original district of Bulungan in late-1999 resulted in the creation of three new districts: Bulungan, Malinau, and Nunukan. At the same time, Kutai was also divided up into three new districts: Kutai Barat, Kutai Kertanegara and Kutai Timur. Post-partition Bulungan (known hereafter as 'Bulungan') encompasses an area covering the entire lower coastline of the original district, while Malinau is land-locked in the region's interior and stretches along much of the original district's western border with the Malaysian state of Sarawak. Malinau shares borders with both Bulungan and Kutai Barat. Kutai Barat is also landlocked and shares a border with Sarawak, along with both the provinces of West Kalimantan and Central Kalimantan. Of the three new districts covered in this survey, Malinau encompasses by far the largest area, covering over 42,000 km^2, followed by 32,000 km^2 for Kutai Barat and 13,000 km^2 for Bulungan, respectively (BPS, 2000 & 2002). All three districts have numerous sub-districts. Malinau is organized into five subdistricts, while Kutai Barat has as many as 15 (Barr et al., 2001; Pemkab Kutai Barat, 2004).
In 2002, there were 19.6 million ha of forest in East Kalimantan. Using classifications developed by central government, 15 million ha (76.5 percent) of this was designated for or had already been set aside for timber production, plantation development or agricultural conversion (BPS, 2002). The remainder was classified as National Parks and Reserves (1.8 million ha) and Protection Forest (2.8 million ha). The huge Kayan Mentarang National Park (1.3 million ha) is contained within Malinau. Official forest cover in all three surveyed districts was still relatively high, with as much as 95 percent of Malinau's land area officially classified as Forest Estate (*Kawasan Hutan*) (Barr et al., 2001). The actual forest cover was probably a lot lower. For example, in Kutai Barat the district government estimated that 50-60 percent of the district was still covered in forest (Casson, 2001).

Figure 2.1 East Kalimantan and Districts

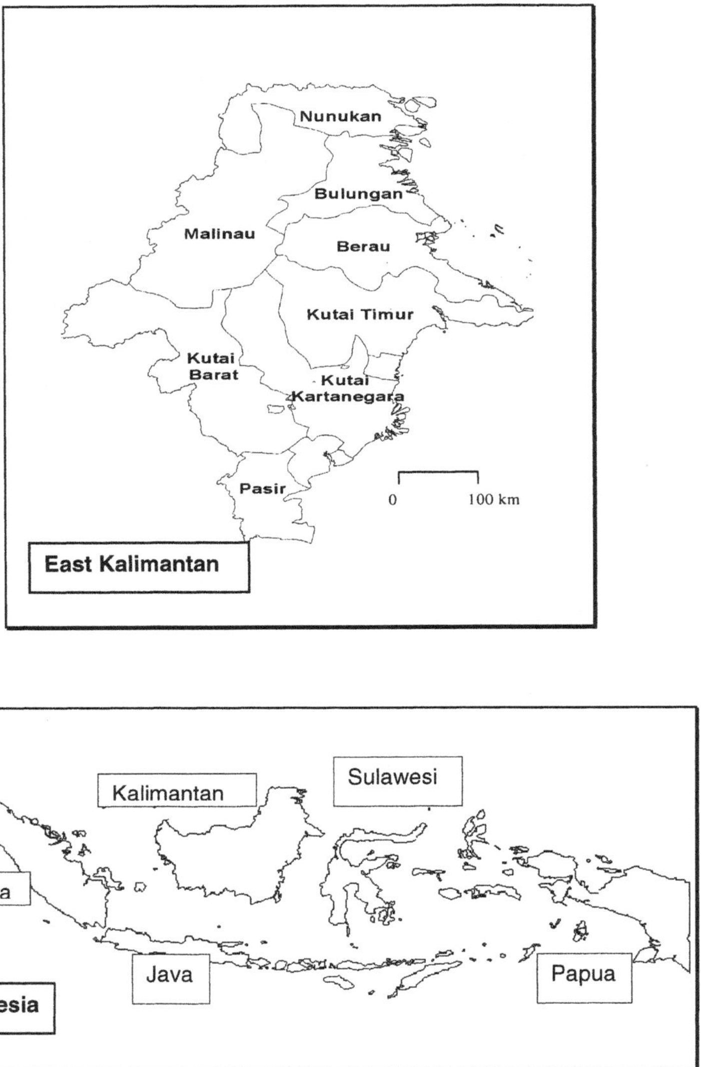

Source: Author's illustration

Thus, all three districts are characterized by heavily forested, and sometimes, mountainous terrain. Paved roads were rare and channels for ground transport through much of the rest of the district tended to be limited to unpaved logging roads. In the more rugged or remote parts of these districts, even these were non-existent. Historically, rivers have served as the main arteries for travel and trade in all regions and in many places were still the main form of transportation for many communities.

2.3.2 Population, People and Livelihoods

In 2002, the population of East Kalimantan province numbered 2.6 million people in a country with an estimated 220 million citizens (BPS, 2002; UNDP, 2003). The province's population has increased by more than 300 percent since the early 1970s, mainly due to intensive in-migration caused by growth in natural resource based industries such as timber, oil, gas and coal mining. The urban areas of Samarinda and Balikpapan as well as adjacent parts of Kutai Kartanegara district, around the lower Mahakam River, were inhabited by 55 percent of the province's population. The rest of the province's area, particularly the hinterland including Malinau and Kutai Barat, has always been sparsely populated due to its inaccessibility.

Malinau's estimated population was just over 35,000 people in 2002, or less than one-half the population of Bulungan with an estimated 85,000 in 1998 (Bappeda Tk II Bulungan, 1998). Kutai Barat had a population of nearly 150,000 people in 2003, living in and around the administrative district capital[18], Melak, and over 200 villages (Pemkab Kutai Barat, 2004). Population densities outside the district capitals in all three districts were very low. For example, most sub-districts in Malinau had a density as low as 1.0 person per km^2 (Barr et al., 2001). In all three districts, a significant proportion of the populations lived in the district capitals. Approximately 20,000 people, or 60 percent of the Malinau's total population, lived in and around Malinau town (Barr et al., 2001). There, the population was composed of a diverse mix of ethnic groups, dominated by the indigenous Dayak Lundaye/Putuk, Punan, and Tidung and with significant numbers of Bugis, Makasarese and Javanese migrants. In the Malinau sub-districts, there were small numbers of Dayak shifting cultivators and Punan hunter-gatherers highly dependent on the forest. In Bulungan, the distribution and composition of the population in the district capital and in outlying areas was similar to Malinau. Indigenous groups included the Kayan, Kenyah, Punan and Brusu groups along with the non-Dayak Bulungan and Tidung people. Kutai Barat was again very diverse, containing groups of Kayan and Ken-

[18] Sendawar is the 'official' district capital of Kutai Barat (Casson, 2001).

yah along with large numbers of Benuaq and Kutai peoples. Similar to Malinau, there were increasing numbers of migrants from other Indonesian islands such as Javanese and Bugis, both in Bulungan and Kutai Barat.

In the communities of all three districts, livelihoods were predominantly based on subsistence agriculture and the collection of non-timber forest products. Subsistence agriculture typically revolved around the planting and harvesting of rice (*sawah*), and non-staple food crops such as chillis and tomatoes (Suramenggala et al., 2001). Other than the sale of non-timber products, e.g. rattan, honey and timber, cash incomes tended to derive from the sale of agricultural surpluses and plantation production and were hence, dependent on the season and the success or otherwise of crops. Formal jobs were sometimes provided by local government, e.g. teaching, and commercial logging companies. In Bulungan and Kutai Barat, there was also employment provided by commercial plantations as well, for example growing coconuts, cocoa and coffee. Communities located nearer to district capitals also had relatively large numbers of people working in other private enterprises such as taxi drivers and limited factory work.

2.3.3 Economy and the Timber Trade

East Kalimantan is officially one of the richest provinces in Indonesia with a GDP per capita of over USD 5,000 in 2002, set against a national average of over USD 800 per capita (UNDP, 2003). This has been mainly due to the abundance of natural resources including timber, oil and gas in the province. For a long time, East Kalimantan has been one of the most important timber producing provinces in the whole of Indonesia, with perhaps as much as 40 percent of Indonesia's log production originating from the province between 1970 and 1979 (Potter, 1990). Forests in East Kalimantan were known for having an exceptionally high density of Meranti species (Ross, 2001). Historically, these hardwood species made up around 85 percent of total wood production with others such as Ulin and Medang comprising the remaining 15 percent (Suramenggala et al., 2001). However, the forestry sector has been declining since the mid-1980s, with natural gas and oil becoming the main contributors to the province's economy instead. In 2002, oil and gas generated about 60 percent of export value in East Kalimantan (BPS, 2000 & 2002). The remaining 40 percent were generated by coal (and other mineral) mining as well as forestry, with coal mining showing particularly strong increases in production in the 1990s.

Commercial logging began in Kutai Barat in the late 1960s, shortly after Suharto's government opened Indonesia's forestry sector to private investment. In the beginning, large-scale HPH concessionaries were primarily located in easily accessible areas along navigable rivers such as the Mahakam in pre-partition Kutai. It was not until the timber boom of the late-1970s that the Forestry Department began distributing HPHs in portions of the region's interior. Between 1970 and 2000, HPHs extracted about 156 million cubic metres (m^3) of logs from the province's forest. At their peak in 1992, active HPHs totalled 89 companies (DKKT, 2000). Since then, the number of active, large logging concessionaries in the province has declined, mainly due to a fall in profitability (Casson and Obidzinski, 2002). As of 1999/2000, there were 83 HPH permits in effect, of which 65 were active and 17 inactive (DKDA, 1999/2000). The effective HPH permits had 8.8 million ha of forest under licence. In 1999, official log production peaked at 7.2 million m^3, although this had fallen to around 4.5 million m^3 in 2001 (DKKT, 2002). Production from IPPK/HPHH concessions was not necessarily recorded in these data (see Obidzinski and Palmer, 2002).

2.3.4 Timber Prices, Timber Markets and Taxes

A significant proportion of timber produced from IPPK concessions in Malinau and Bulungan was processed in the sawn wood and plywood mills of Sabah, Malaysia, in addition to being 're-exported' to other markets (see Palmer and Obidzinski, 2002; Obidzinski and Palmer, 2002). This trade continued after a timber export ban was imposed by the Indonesian government in October 2001. Table 2.1 shows the prices of one of the most common species of timber found in East Kalimantan and illustrates the price incentive for 'illegal' exports from Malinau and Bulungan.

Table 2.1 Price Comparison of Red Meranti Logs in Indonesia and Malaysia, 1999-2002

Country	1999	2000	2001	2002*
Indonesia	60-70	60	30-35	45-50
Malaysia	115-125	110-115	80-95	100-105

Source Province and district forestry offices in East Kalimantan; Sabah state forestry offices; timber traders (from Palmer and Obidzinski, 2002).

Notes All prices in USD; * Until April-May

Thus, domestic prices only fetched around half of the price to be found in Malaysia. This, coupled with the relative lack of domestic processing timber plants found in Malinau and Bulungan and in the north of East Kalimantan generally and the proximity and ease of access to Malaysian timber markets resulted in a steady source of demand for timber from small-scale concessions in Sabah (Obidzinski and Palmer, 2002). In Kutai Barat further south, much of the timber produced from HPHH concessions was processed among the multitude of sawnwood and plywood mills located hundreds of kilometres downstream on the banks of the Mahakam river towards the provincial capital, Samarinda. Thus, timber from these concessions most likely fetched prices in the range of USD 50 per m^3.

IPPK permit-holders were not required to pay any national and provincial-level royalties and there were no reforestation or selective logging requirements associated with the HPH concession system, although they were subject to fees imposed by district governments (see table 2.2). The district government of Malinau generated formal revenues from the allocation of IPPK permits through the payment of two fees (Barr et al., 2001). First, IPPK-holders were required to pay a one-time 'third party donation' (*sumbangan pihak ketiga*) of Rp. 200,000 (USD 22.22)[19] per ha. This was effectively the cost of applying for and holding an IPPK permit, and was similar in Bulungan. The second fee that recipients of IPPK permits were required to pay in Malinau was a 'production retribution' (*retribusi produksi*) of Rp. 15,000 (USD 1.67) per m^3 of timber harvested[20], a figure that rose to Rp. 60,000 (USD 6.67) per m^3 in Bulungan. District governments in Bulungan and Malinau both charged an IPPK export tax of around Rp. 150,000 (USD 16.67) per m^3, which was officially suspended when the timber export ban was imposed.

At a maximum size of 10,000 ha, a HPHH concession could be potentially much bigger than IPPK but was apparently a more complicated affair due to requiring slightly more investment and bureaucracy (Palmer and Obidzinski, 2002). This was because HPHH operators were supposed to follow many of the rules followed by HPH concessionaires, including a 'reforestation tax', and a tax to the central government in addition to a production tax to the district government. There were no export taxes and in some cases the reforestation tax was paid directly to communities (see chapter 10). Similar to the IPPK system, the district government in Kutai Barat charged fees for the permit application process, upwards of Rp. 20 million (USD 2,222.22) for a HPHH of 100 ha. This works out at around the same as the fees paid in Malinau and Bulungan.

Overall, the cost bases for IPPK and HPHH concessionaires were broadly similar, which was mainly because tax avoidance and informal payments were so widespread (see Smith and Obidzinski, 2002; Palmer and Obidzinski, 2002). For example, Suramenggala et al. (2001) describe the payments made in Bulungan. All levels of government, including the district head and sub-district

---

19 'Rp.' and 'IDR' both denote the currency used in Indonesia, the Rupiah, and are used interchangeably in this study. Unless noted otherwise, the exchange rate used for all conversions in this study is USD 9,000 = IDR 1.00.

20 With these two fees, Barr et al. (2001) projected that the district government should have been able to collect approximately Rp. 53 billion in revenues for the district budget from IPPKs issued through February 2001 (or USD 5.3 million at an exchange rate of Rp. 10,000 per USD). This amounted to roughly nine times Malinau's planned budget for 2000.

heads along with the army and police extracted some form of payment or 'wild tax' (*pungutan liar*) from the permit holders. While the actual nature and size of these informal payments varied, the overall cost of setting up an IPPK concession of around 1,500-2,000 ha was approximately USD 200,000 according to Palmer and Obidzinski (2002). Costs included heavy equipment, logistics and basic logging roads, along with up-front official and unofficial payments. Once production began, informal payments continued, although it seems that illegal exports required higher absolute amounts in informal payments than that transported within East Kalimantan. Despite this, profitability was apparently not significantly impacted (Obidzinski and Palmer, 2002). It seems likely that higher timber prices in Sabah more than offset the increased demand for bribes following the imposition of the log export ban in Indonesia. Thus, an IPPK/HPHH concessionaire stood to make a profit of perhaps up to USD 40 per m^3, which for a concession of 2,000 ha with a low rate of harvesting of say 20 m^3 per ha might have made profits of over USD 1.5 million. This easily covers start-up capital costs. A higher harvest rate of say 80 m^3 per ha would have resulted in a four-fold increase in profits, to around USD 6 million.

Table 2.2 Comparative Profitability of HPH, HPHH, IPPK Concessions and Illegal Logging (all in USD)

	HPH	HPHH	IPPK	'Illegal Logging'
Timber (log) prices per cubic meter	100	100	100	55
Tax obligations:				
Felling/extraction/transport permits	Yes	Yes	Yes	No
Reforestation	18	18	No	No
Central government tax	7	7	No	No
District income tax (*Retribusi*)	No	3	3-6	No
District export tax	No	No	15	No
Community benefits	?	0.05-15	0.05-15	?
Informal payments	10-20	10-20	10-20	20-30
Extraction	20	16	16	5
Total Cost	65	64	54	23
Net revenue	**35**	**36**	**46**	**32**
Investment	Large	Moderate	Small	Very small
Bureaucracy	Extensive	Limited	Limited	None

Source District and province forestry offices, HPH concessionaries, IPPK operators and timber brokers in the area (adapted from Palmer and Obidzinski, 2002).

Notes 1. All figures in USD; 2. District export tax is based on export regulations in Bulungan that became effective in April 2001 but was suspended in October the same year as result of the log export ban.

3 Literature Review and Conceptual Framework

This chapter reviews the concepts that are used as the theoretical basis for the empirical analysis in Part III. First, the political economy literature on the impacts of decentralization is presented, followed by the conceptual framework for the assessment of impacts of decentralization on Indonesian communities. The contribution to the literature is given along with the expected direction of effects to be empirically tested in chapter 11, followed by a presentation of the model used to assess the underlying determinants of the variation in outcomes from post-decentralization logging agreements. Again, the contribution to the literature is presented along with the hypotheses to be tested in chapter 12.

3.1 Comparison of the Impacts of Mechanised Logging on Communities Before and After Decentralization

3.1.1 The Political Economy of Decentralization

Many countries around the world have adopted some form of decentralized governance (Agrawal and Ribot, 1999), and over 60 have decentralized some aspect of natural resource management (Agrawal, 2001a). Decentralization is typically referred to as a transfer of powers from central government to lower levels in a political-administrative and territorial hierarchy (Agrawal and Ribot, 1999). The main arguments for decentralization are that it can be a way of increasing both efficiency and equity in natural resource management (Ribot, 2002). It may also be a way of leveraging the development of local democracy: 'the efficiency and equity benefits from decentralization come from the presence of democratic processes that encourage local authorities to serve the needs and desires of their constituents' (Ribot, 2004:2). Decentralisation as defined here usually implies a top-down process, although development that includes local empowerment depends on bottom-up processes (Chambers, 1997).

Thus, decentralization can be used as a tool for promoting development, with the main aims of increasing efficiency, equity and democracy (Larson, 2004). Efficiency outcomes are not, however, a focus of this study. Of greatest concern to local communities, are equity and democracy considerations, specifically, a greater control over livelihoods and a greater share of natural resource benefits (Edmonds et al., 2003). Ostrom (1990) argues that local participation in decision-making results in a greater sense of ownership over these decisions, such as rules for resource use. This should, in theory, result in an increased willingness to abide by them, hence increasing their effectiveness (Carney, 1995). In contrast, distant state authorities are less effective at allocating forest resource rights than local authorities simply due to the vast size of forests under their control (Carney and Farrington, 1998). Due to this sense of 'ownership', the onus is on communities to be more engaged in implementing, monitoring and enforcing logging agreements. Participation and equity considerations from decentralization and IPPK/HPHH concessions are further discussed in chapters 7, 10 and 13.

Community-level management can have ecologically and socially positive outcomes (Ribot, 2004). Larson and Ribot (2004) document cases of decentralized natural resources that have been effectively managed by local communities. However, in a review of forest sector decentralizations in 20 developing countries including Indonesia, Larson (2004) demonstrates that true, democratic decentralization has only rarely been implemented. Substantial decision-making power, resources and benefits from forests are still centralised in many countries. Even in 'truly' decentralized cases, outcomes vary among countries and regions, although, there is little evidence that decentralization has benefited forests and the people who depend on them (Kaimowitz et al., 1998). Nevertheless, the degree of decentralization and its outcomes are dependent on many factors including political and social histories, a failure to account for time and insecurities and unresolved land and forest tenure relations (Larson and Ribot, 2004: 8).

'Big Bang' decentralization moved Indonesia from being one of the most centralized countries in the world to one of its most decentralized (Hofman and Kaiser, 2002). As detailed in chapter 2, *de facto* decentralization followed by imperfect decentralization reforms in the forest sector, resulted in communities exerting property rights over customary forest. From 1999 until 2003, many communities engaged in direct negotiations and made legalized agreements with logging companies in exchange for access to financial and non-monetary benefits. Casson and Obidzinski (2002) concluded that local people have benefited, at least financially and in the short-term, from the post-decentralization, mechanized concessions regime compared with the situation before decentralization. Studies such as Resosudarmo (2004) on the other hand consider the effects of decentralized forest management including ecological impacts to be undesirable, and preliminary impacts specifically on communities are considered to be 'mixed' (p.126). Despite that, there have been some positive effects in addition to retaining a higher proportion of timber rent, at least compared with the period prior to decentralization; local people now have greater access to forest resources and have increased formal recognition of forest claims. Edmonds and Wollenberg (2003) stress, however, that in order to analyse the impacts of decentralisation, it is important to understand what existed previously in order to place the impacts on local people in context.

A conceptual framework is presented to answer the question of whether or not there have been any significant differences between the impacts of centralized and decentralized systems of mechanized logging concessions on local communities. Five categories of impact are presented: financial, enforcement, rent-seeking, social and environmental. Given *de facto* decentralization and to avoid confusion, the words 'decentralization' and 'post-decentralization' will refer to a time period including *reformasi* as well as the formal implementation of decentralization laws, i.e. including all events from 1997-98 until the present day; 'pre-decentralization' will refer to the period before 1997-98, and specifically the Suharto era (1966-98).

3.1.2 Conceptual Framework

3.1.2.1 Financial Impacts (Monetary and Non-monetary Benefits)

Until 1997-98, the financial benefits from large-scale, mechanised logging and other forest industries flowed largely to a small elite consisting of the Suharto family and their patrons in the government. Casson and Obidzinski (2002) state that local people living in, or around, HPH concessions received 'very little' from the companies before decentralization (p. 2135). Thus, HPH companies generally provided no cash and few, if any social developments to communities at this time, despite government regulations mandating that they set up rural development programmes. Furthermore, in many places villagers even had to ask for permission from HPH companies before they could gather forest products or harvest wood for the construction of houses or village buildings (Sudana, 2004). Hence, some communities still *de facto* used forests for agriculture, hunting, fuel and other products, often with the approval of the HPH (see Anau et al., 2002).

After decentralization, in exchange for providing access to commercially valuable timber, communities typically negotiated for and expected to receive production-based fees per m^3 of timber produced, and non-monetary benefits (Casson and Obidzinski 2002; Barr et al., 2001). Promises of non-monetary benefits varied from the building of schools and churches to rattan gardens and seeds for plantations. While the promised fees were not necessarily the fees that communities actually received, Casson and Obidzinski (2002) note that they were 'much more than local people were able to obtain during the Suharto era' (p. 2142). The IPPK/HPHH permit system was supposed to generate some employment for communities, although companies preferred to bring in their own, experienced 'crews'; this may also have been a strategy to prevent communities from striking against logging activities (see Palmer, 2004). Logging activities may have had positive economic impacts on other, local industries, e.g. sawmills, at least in the short-term (Casson and Obidzinski, 2002).

There is a need to distinguish between negotiated and actual payments, which could be potentially quite different due to enforcement problems and the appropriate measure to be used here is the actual payments received by communities. As table 3.1 shows, significant differences are expected between financial and non-monetary benefits flowing to communities when comparing the pre-

and post-decentralization regimes: zero financial benefits and zero to 'some' non-monetary benefits before decentralization (this benefit range denoted by '0/+'), and; zero to 'some' financial and non-monetary benefits after decentralization (again denoted by '0/+'). Hence, it is expected that:

- Financial and non-monetary benefits to communities increased with decentralization (I).

Empirically, financial and non-monetary benefits received can be observed on the basis of absolute values of monies received plus the value of social developments (for the whole community). Social developments are valued on the crude basis of the cost of provision only due to the variation in these and a lack of detail necessary to value them more precisely as a flow of future benefits. Employment in the community, where provided, can be valued on the basis of the number of employees, the wage rate and the length of employment. Total values of benefits provided can also be broken down and estimated on a per person or per household basis. Alternatively, the proportion of households that received financial benefits and/or social developments provided by the company could proxy for benefits received in the absence of precise data.

3.1.2.2 Social Impacts

3.1.2.2 i. Empowerment

Similar to the non-mechanized small-scale concessions of 1967-71, district governments post-decentralization issued small-scale permits that were supposed to benefit local communities. The crucial difference is that the small-scale concessions in the decentralization era all employed mechanised methods of exploitation and operated on a much larger scale than the older, non-mechanised concessions. However, the need to collaborate with brokers and outside logging contractors led Resosudarmo (2004) to argue that the need for capital and mechanisation has resulted in a greater share of revenue remaining outside the communities and less control and empowerment at the local level than people were expecting. Nevertheless, these timber networks tapped into communities' desire for control and empowerment and the newly found ability of community leaders to openly and independently make their own decisions regarding forest usage.

Table 3.1 Impacts of Mechanised Logging on Communities: A Framework for Analysis

Type of impact on community			Empirical indicators in survey results	Expected direction of effect of decentralization	
				Pre-	**Post-**
Benefits	Financial (I)		Total value in IDR per community, IDR per household, proportion of households receiving fee	0	0/+
	Non-monetary (I)		Type and total value in IDR per community, proportion of households aware of benefits, proportion of households with company employment, wage rate (IDR per day), length of logging operation (days)	0/+	0/+
	Social ('empowerment') (II)		Proportion of households indicating perception of forest ownership	0	0/+
Costs	Social ('trust')		None	0	0/-
	Enforcement (III)	Firm (IIIa)	Proportion of households participating in activity against firm, type of activity, length of conflict (days)	0	0/-
		Other community (IIIb)	Proportion of households indicating conflict from forest border/claims	0	0/-
	Rent-seeking (cost of unproductive activity) (IV)		Proportion of households indicating problems over distribution of financial and non-monetary benefits	0	0/-
	Ecological (local use values) (V)		Proportion of households indicating negative effects on water quality, flooding, hunting, forest products, farming from logging	0/-	0/-
Overall (most positive outcome/most negative)				**+/-**	**+++/-----**

Note IDR denotes Indonesian Rupiah; '+' denotes positive outcome, '-' negative outcome.

Thus, direct negotiations and agreements between communities and companies were a tangible acknowledgement of customary rights and a signal of community empowerment, free from central government control. An attempt to measure this sense of empowerment is made by studying any changes in perceptions of forest ownership from the pre- to the post-decentralization periods. It is expected that:

- The perception of community forest ownership and, in this sense, empowerment, increased with decentralization (II).

3.1.2.2 ii. Social Costs from Rent-seeking and Conflict

Intra-community rent-seeking (see 3.1.2.4) affects equity among community members, and may thereby lead to intra-community conflicts. The concept of equity is very strong among traditional, forest-dependent communities, irrespective of religion[21]. Thus, internal divisions and conflicts, unless quickly and amicably resolved could lead to longer-lasting impacts on social capital, i.e. a reduction in cohesiveness, trust and the ability for collective action. Following Hoddinott (2002) who states that where there is weak social capital, community participation may lead to elite rent capture, it is possible that this may result in a vicious circle of rent-seeking and social capital erosion. Moreover, there may be social costs associated with inter-community conflicts over forest claims (see 3.1.2.3) that might affect community relations and future cooperation. While it is acknowledged that such costs might arise both in the short and the long-term, an explicit quantification of these was unfortunately outside the scope of this study. All that can be inferred is that there may be a higher likelihood of negative social impacts in communities where there are higher intensities of unresolved intra- and inter-community conflicts.

3.1.2.3 Enforcement Costs

3.1.2.3 i. Enforcement vis-à-vis the Company

For Malinau district, Sudana (2004) describes how conflicts between communities and HPH companies did not begin until *de facto* decentralization began in 1997-98, when people were becoming more aware of the economic and political realities of the period. Prior to this period, conflicts were expressed through complaints and community meetings. Direct complaints or action against companies were relatively rare because people were afraid of them. In the Suharto

[21] In East Kalimantan, local communities tended to be either Muslim or Christian, sometimes in combination with animalist traditions. It should be noted that there is a phenomenon known as 'identity card Muslim' in Indonesia. By law, people have to announce a religious affiliation in order to claim state benefits.

era, when communities did protest against HPH companies, the police and military tended to intervene on the side of the company, and conflicts were sometimes dealt with rapidly and violently (Makinuddin, 2001; GFW/FWI, 2002).

As noted by many authors (e.g. Casson and Obidzinski, 2002), communities were vulnerable to companies' failure to comply with IPPK/HPHH agreements. The government's capacity for enforcement was very weak (Richards et al., 2003), and weakly defined property rights imply that rights to the forest could be claimed either by communities or companies (Engel, López and Palmer, 2006). In the case of the company this usually took the form of making promises that were not complied with later (Barr et al., 2001; Palmer, 2004).

Communities learnt very quickly how to deal with IPPK/HPHH companies, particularly after enforcement attempts through dialogue with company officials or direct reporting to the local government yielded relatively little in the way of results (Limberg, 2004). Communities sometimes took direct action such as demonstrations and road blocks against non-complying companies when other conflict resolution mechanisms had already been exhausted, i.e. as a last resort. The success of community self-enforcement actions may have depended on a community's ability for collective action (Palmer, 2004; Engel and Palmer, in press). Community blockading actions drawn out over a longer period of time would need greater collective action within the community and hence a higher level of household participation, support and organization in order for them to be sustained. Coordination and leadership was usually, but not always provided by village leaders such as the village head and customary law head. Thus, overall, communities found agreements difficult to monitor and enforce. Direct methods and strategies to enforce agreements such as demonstrations and road blocks imply relatively high opportunity costs of time and transaction costs for communities. Sudana (2004) notes how these conflicts between communities and companies in Malinau were mostly settled by cash compensation. It is expected that:

- Enforcement costs of blockading to communities increased with decentralization (IIIa).

Empirically, community-company conflicts can be observed on the basis of the proportion of households in the community participating in an enforcement activity, the nature of the activity, the amount of time the activity consumed and whether or not it successfully stopped company operations.

3.1.2.3 ii. Enforcement vis-à-vis Other Communities

Decentralizations lead to a redistribution of rights over and benefits from resources that result in some benefiting more than others (Ribot, 2004). In Indonesia, economic incentives from community-company agreements have intensified conflicts among communities, such as disputes over village borders. As Barr et al., (2001) and Casson (2001) note in Malinau and Kutai Barat, respectively, the availability of IPPK permits has dramatically increased competition regarding resource claims given the weak legal basis for village boundaries or rights to forest resources. Sudana (2004) also describes how the increase in post-decentralization forest-related conflicts among communities in Malinau, while rooted in historical relationships among local settlements and ethnic groups in the area, was at least partly influenced by the proliferation in IPPK agreements. Casson and Obidzinski (2002) note that as more communities entered into these agreements, they became more vigilant about delineating village forest borders and enforcing these.

Typically an inter-community conflict involved forest border claims and the logging company working on behalf of, and benefiting one of the communities. For example, Iwan (2004) describes how logging operations from at least two different communities spilled over a disputed border into the territory of Setulang community in Malinau district. In both cases, Setulang enforced its property rights over the forest through the confiscation of logging equipment and demands for compensation were made for the harvesting of timber in Setulang's territory. Similar to company non-compliance and community-company conflicts, a community would convey its complaints and demands with letters and personal visits to company and government offices, and then if necessary, undertake direct actions such as demonstrations (Sudana, 2004). Matters would then either be resolved through negotiation or the conflict would continue, although these kinds of conflicts tend not to be resolved so easily, with possible long-term consequences for inter-community relations.

With respect to inter-community conflicts, while it is acknowledged that many have historical origins, these were not necessarily the product of mechanised logging on community territory prior to decentralization. Therefore, there was no enforcement cost from these conflicts at that time (denoted by '0' in table 3.1). It is expected that:

- Enforcement costs from inter-community conflicts increased with decentralization (IIIb).

Empirically, inter-community conflicts can be observed on the basis of the proportion of households citing problems with neighbouring communities, the nature of the problem and whether or not it derived from post-decentralization concessions.

3.1.2.4 Rent-seeking Impacts

There is anecdotal evidence that some groups undertook action against the company independent of others within the same community, even when the company had complied (at least partially) with the agreement (Limberg, pers. comm., 2003). These actions were generally short-lived and the payoffs relatively small. The individual groupings were simply seeking extra rents for themselves, which could be interpreted as rent-seeking rather than self-enforcement behaviour. A single roadblock manned by just a few people would be enough to stop a company operating, at least in the short-run. In addition, this kind of rent-seeking by community members (known hereafter as 'opportunistic blockading') may also give the firm an increased disincentive to comply further with the agreement made with the community. Thus, 'opportunistic blockading' could lead to intra-community problems, which may also potentially undermine the community's capacity for collective action and self-enforcement of the agreement should the company subsequently fail to comply anyway.

In countries where there is serious corruption at both the local and national level, decentralization may allow more local elites to participate in rent-seeking activities (Larson, 2004). Rent seeking by actors at the district level has intensified since decentralization with the increasing political power of district heads (Palmer and Obidzinski, 2002). However, rent-seeking by national elites may decline with decentralization, so the net effect is unclear. Should decentralization lead to the capture of rents by local elites, this may stymie the equitable distribution of these rents (Carney, 1995). Following deals made with logging companies, the culture of rent-seeking in Indonesia has spread from well-established local timber networks consisting of members of district government and business elites, to the cooptation of village elites. A number of case studies from East Kalimantan show evidence of widespread rent-seeking and rent capture by well placed village elites that failed to distribute logging fees paid by companies[22] (e.g. Barr et al., 2001; Limberg, 2004). This is despite the fact that many communities had already agreed on systems of rent distribution.

Rent-seeking by elites took two forms. First, the company may have complied with at least some of its payments, which were not subsequently shared or distributed by village leaders. Instead, these were captured by the leaders who then blamed the company for non-compliance when villagers complained about not receiving their share of payments. Hence, community-company conflicts might have occurred even when the company had complied with at least part of the agreement (hereafter denoted as 'useless blockading'). At some stage though, people usually realised that their leaders were capturing and not sharing rents, and instead of blaming the company, would accuse the leaders of wrong-doing. The second form of elite rent-seeking involved leaders taking 'special fees' from the company in exchange for allowing the company to avoid compliance with some or all of its contractual obligations. In exchange, the company

[22] Hoddinott (2002) argues there is an even higher risk of rent capture of benefits by local elites to the detriment of the poor in fractionalised communities where trust and/or social capital is weak.

would expect the village leaders to promise not to coordinate or dampen the effects of any actions against the company should people get angry about firm non-compliance, i.e. elites would try to undermine community's willingness and ability to self-enforce contracts (e.g. see Suramenggala et al., 2001). In either case, intra-community problems are expected to occur where rent-seeking by elites has taken place, and as before, these problems are expected to lead to less incentive for community collective action should the company then decide not to comply further with its contractual obligations.

In this framework, it should be noted that while local, intra-community rent-seeking problems are not necessarily caused directly by the behaviour of logging companies, they are related to the responsibilities of communities in making these agreements and as such did not occur prior to 1997-98. Thus, in this sense they are an impact of decentralization; people use up resources in order to seek rents[23]. Hence, rent-seeking can be considered an 'unproductive activity' (Pearce, 2000). It is expected that:

- Unproductive activity costs to communities resulting from rent-seeking increased with decentralization (IV).

To empirically distinguish 'true enforcement' activities against firms from 'opportunistic blockading' and 'useless blockading', two criteria are adopted:

- *Occurrence of non-compliance by the company.* If there is no evidence for non-compliance but still blockading by the community, then it cannot be 'true self-enforcement';
- *Intensity of intra-community conflict.* If there is a high intensity of intra-community conflict, then this is more likely to be due to opportunistic or useless blockading, hence indicating rent-seeking within the community.

3.1.2.5 Environmental Impacts

Mechanized logging requires the building of roads through the forests and the use of heavy equipment, typically leading to widespread soil degradation and damage to uncut trees (Ross, 2001). According to Ross, before decentralization, the Ministry of Forestry barely regulated HPH concessions and concessionaires apparently violated the reforestation regulations with impunity. Casson and Obidzinski (2002) describe how, despite controlling extensive areas of forest, the HPH companies were long harvesting timber over the Ministry of Forestry's maximum approved level and obtained timber from 'illegal' sources in order to meet demand. For example, Schwarz (1990) estimated that HPHs in the 1980s

---

[23] While there were almost certainly rent-seeking costs before decentralization, particularly at the national level (see for example Palmer, 2001), this analysis focuses on costs incurred at the local level only.

were illegally removing around two million m^3 of timber from protected areas every year. As a result, forests were logged in such an unsustainable manner that it often caused erosion, flooding, and drought (GWF/FWI, 2002). These environmental consequences have not taken very long to impact on local communities. For example, in 1998, degraded watershed forests in Aceh, heavily logged in the 1990s, were no longer carrying out their hydrological functions. After heavy rains, streams quickly flooded and then tended to dry up afterwards (McCarthy, 2000). Thus it seems that the exclusion of communities from access to forest resources, in order to promote commercial logging, has led the same people then having to live with the effects of related degradation (Edmonds et al., 2003).

After decentralization, large numbers of IPPK/HPHH permits were issued prior to the establishment of a regulatory agency at the local level, along with insufficient capacity to monitor the implementation of licenses (Barr et al., 2001; Casson, 2001). Moreover, there were indications that permits were approved by district governments on the basis of informal incentives such as bribes rather than environmental or any other considerations. Thus, local governments failed to regulate or monitor the environmental performance of IPPK/HPHH operations, such as selective cutting or other harvesting regulations or the enforcement of protected land designations (Casson and Obidzinski, 2002; Resosudarmo, 2004). Resosudarmo (2004) notes that while it might be still too early to draw conclusions on the environmental impacts of post-decentralization small-scale concessions, the changes so far indicated 'a substantial increase in logging with little regard for environmental consequences....This increase is likely to lead to forest deterioration and conversion.' (p. 113)[24]. Limberg (2004) by contrast argues that the impacts of two years of IPPK operations on the forest in Malinau were relatively smaller than initially feared, although he attributes this more to luck than well-designed and implemented forest management policy. Along with the failures of local government, the central government has been unable to enforce controls as it did during the Suharto period (Casson and Obidzinski, 2002).

Wollenberg (2004) reports on a quantitative study undertaken by Iskandar and his colleagues, which compared the impacts of two HPH concessionaires with three IPPK permit holders. They found that the IPPK holders harvested more intensively, removing four times as many trees per unit area and harvesting

---

[24] Resosudarmo (2004) also describes how IPPK/HPHH led to an increase in the area logged and volume extracted, although without providing any supporting data. The Indonesian Ministry of Forestry (2004) has provided some data although there are doubts about the accuracy and reliability of government data (see Obidzinski and Palmer, 2002). There are also problems of interpreting the relatively limited satellite data available on Indonesian forests. For example, Limberg (2004) reports on satellite evidence showing that the actual IPPK area logged was less than the area stated in small-scale permits for the Malinau watershed.

a higher proportion of younger trees. IPPKs also caused significantly more damage to residual trees and larger canopy openings, resulting in a more degraded and fragmented forest. This damage further threatened the potential for forest regeneration. Consistent with Iskander et al.'s findings, Dudley (2001) asserts that given political and economic uncertainties, the effect of the IPPK/HPHH system on the forest may be worse than the HPH system. Moreover, many IPPKs operated in old, logged-over HPH areas (see Barr et al., 2001), which is likely to cause more damages (Scotland et al., 1999).

On the other hand, Barr et al. (2001) concluded that in reality, the pattern of forest degradation is no different from what it was under centralised management; in both cases there was little concern in or about long-term sustainability. Overall, it seems that the environmental impacts from IPPK/HPHH concessions were at the minimum as bad as, and possibly worse than impacts under the old, centralised system especially for those concessions situated in recently logged-over areas. Table 3.1 captures some of these costs as ecological, local use values[25]. Following Barr et al. (2001), the expected direction of effect in the short-term ranges from zero to negative (denoted as '0/-'). It is expected that:

- Ecological costs to communities neither increased nor declined with decentralization (V).

Empirically, perceptions of environmental damages can be elicited by asking about whether or not there have been any changes in selected environmental indicators, both before and after decentralization, and why these changes might have occurred.

While it is clear that district governments did not and perhaps could not take responsibility for the regulation or monitoring of the environmental performance of small-scale concessions, there has been relatively little research on the role of communities: have they been able to affect any change in the pattern of forest degradation from mechanised logging at all? Dudley (2001) and Resosudarmo (2004) contend that communities were more concerned with immediate economic need rather than long-term forest management. This may be because communities had little confidence that their new, local rights would last, especially given efforts by central government to re-concentrate some of its powers. However, some communities have been observed to shape their agreements to minimise some of the local impacts, for example by negotiating for the location of concession areas along with replanting requirements and the tree species to be harvested (see Palmer, 2004). These observations will be examined further in chapter 11.

[25] Community non-use values, in addition to other non-use, global values are not considered due to being outside the scope of this study.

3.2 Determinants of Community Benefits from Post-decentralization Negotiations

3.2.1 Introduction to the Modelling Approach

Case studies (e.g. Barr et al., 2001, Palmer, 2004) indicate that the outcomes of community-company negotiations after decentralization varied significantly across communities. In some cases communities were able to enforce their informal rights and, hence, effectively bargain with firms, while in others communities appeared to be exploited by firms. Even where communities have been successful in enforcing their rights sufficiently to force firms to actually negotiate, the financial and environmental contract provisions have significantly varied. A game-theoretic model was developed by Engel, López and Palmer (2006) to illustrate the strategic interactions between communities and companies. While this model was motivated by the Indonesian context, it is likely to be relevant to any situation where communities negotiate contracts over natural resource use with outside actors in a context of weak property rights to the resource, a situation frequently found in developing countries (Feder and Feeney, 1991; Alston, Libecap and Mueller, 1999).

The formal literature on the effect of decentralization on natural resource management focuses largely on the determinants of successful collective action by local communities to achieve efficient management outcomes (e.g., Baland and Platteau, 1997a,b, 1998; Ostrom, 1990; Bardhan, 1993a,b)[26]. It generally takes property rights as given and ignores potential interactions with outside actors. Another strand of the literature models conflicts over property rights (Alston, Libecap and Mueller, 1999; Angelson, 2001; Burton, 2004), but does not link potential conflict to the alternative of negotiation. Engel, López and Palmer (2006) model the interaction of local communities with logging companies in a context of uncertain property rights. Property right enforcement is modelled endogenously and linked explicitly to the feasibility and results of community-firm bargaining[27]. To derive hypotheses on the factors influencing the actual payments received by local communities, a conceptual framework is presented based on a simplified version of the models developed by Engel, López and Palmer (2006) and Engel and López (2004). The models presented in these two papers are more general than the version presented here in that they

[26] For a review of this literature see Baland and Platteau (1996) and Agrawal (2001b).

[27] Moreover, Engel, López and Palmer (2006) model the effects of outside intervention by NGOs or other third parties to improve communities' bargaining positions in a situation where at least some of the environmental impacts are not represented by any of the parties in the bargaining process. This is not considered here.

also include the choice of logging area and the possibility that logging area may affect reservation utilities.

The model links conflict and bargaining theory as follows. A major determinant of bargaining positions are the players' reservation utilities. However, in the context of weak property rights, where property rights are often not formally defined and government enforcement is weak, reservation utilities are the outcome of a latent conflict over *de facto* property rights. The model can be interpreted as having two stages. Working backwards, the outcome of what happens during the formation of property rights is presented first followed by community-company bargaining over a mutually agreed logging agreement. The model is used to derive hypotheses on the factors that may be significant in determining the probability that a community can enforce property rights over the forest and on those that determine actual levels of benefits for those communities that have successfully enforced their property rights. Next, hypotheses are derived from the expected total effect of the determinants on property rights formation and community-company bargaining. Hypotheses are tested using econometric analysis in chapter 12.

3.2.2 Property Rights Formation

In Engel, López and Palmer (2006), *de facto* property rights are modelled as the outcome of a war of attrition. This stage adapts ideas from conflict theory and uses them to determine the feasibility of bargaining as well as key parameters affecting it. Conflict theory (see, for example, Dixit and Nalebuff, 1991) usually assumes the existence of two actors.[28] In this case one of the actors has the ability to exploit a resource while the other one may under some circumstances prevent such exploitation. Logging requires a specific factor (e.g., capital) that is available to the logging company (referred to as 'the firm' hereafter), but not to the community. This assumption can be justified on the basis that communities are poor (have low savings) and have a disadvantage in the credit market *vis-à-vis* the firm, associated with capital market imperfections (see for example, Bose, 1998).

The possibility of bargaining arises from the complementarity between the firm and the community in terms of access to the factors of production required for logging. The firm has access to capital, while the community may be able to control access to the natural resource. There are two possible outputs here: timber, the extraction of which requires both capital and the natural resource, and non-timber products provided by the standing forest (e.g., water retention ser-

---

[28] For specific examples of conflict models see Burton (2004), Alston et al. (1999), and Angelsen (2001).

vices, flood prevention, firewood, wildlife), which do not require the use of capital. When the incentives are such that the benefits from the standing forest dominate those of timber products under all feasible distributions, then the community does not bargain and thus the standing forest is protected. Otherwise, and if the community is able to exert property rights, bargaining would be the outcome.

Under the assumption of perfect information, one of the actors can in principle impose its conditions, e.g. the firm may unilaterally exploit the resource if it has enough power to win a war of attrition, or the community may prevent that if the power conditions are reversed. Engel, López, and Palmer (2006) argue that the outcome of this process is asymmetric: if the firm is able to win a war of attrition, the community will effectively lose property rights for the resource and hence, the firm will exploit it unilaterally as an open access resource. If, however, the community is able to win a potential conflict, then it effectively is able to exert property rights upon the resource. That is, the community has formal legal rights—however vaguely defined this might be by law—as well as the ability of effective enforcement. In this case since the community does not have access to capital to enable unilateral exploitation, bargaining between the community and the firm may take place. Moreover, in the perfect information case, conflict is virtual, i.e., players can predict the outcome of the conflict, and the player that would lose the conflict withdraws immediately. With imperfect information, actual conflict is possible, but the outcome will generally depend on the same parameters listed here (see, for example, Burton (2004) for a model with imperfect information).

In addition to assuming perfect information, the following assumptions are made. A firm can attempt to log unilaterally without community consent (resulting in a per-period cost of c), and the community can prevent such unilateral logging by setting up road blocks or sabotaging logging equipment (resulting in a per-period cost of s). Each period the community is able to stop firm operations it obtains use and nonuse value of the undisturbed forest, denoted by b. If, however, the firm wins the conflict it receives profits v from logging unilaterally. If the community wins the conflict it obtains b in every period forever after. The discount rates of the community and the firm are, respectively, denoted by r^C and r^F. The war of attrition is won by the party that is able to stay in a potential conflict longer. Thus, both players can attempt to obtain *de facto* property rights over the resource: the firm by trying to log unilaterally, and the community by setting up blockades to stop the firm from logging. The results of the conflict game are presented in Figure 3.1.

Figure 3.1 Outcomes of the Conflict Game

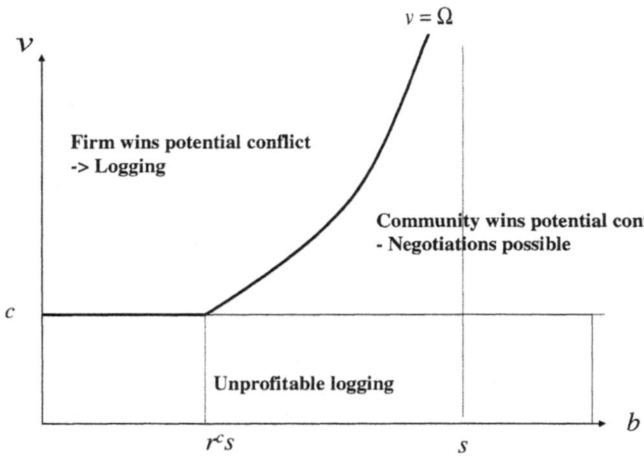

where

$$\Omega \equiv \frac{c}{r^F}\left[\left(\frac{s(1-r^C)}{s-b}\right)^{\frac{\ln(1-r^F)}{\ln(1-r^C)}} - 1 + r^F\right]$$

Source: Adapted from Engel, López, and Palmer (2006)

First, consider the case where $s > b/r^C$. Thus, the costs of setting up a blockade for even just one period already exceed the present value of benefits obtained from the standing forest forever. In this case, the community will never fight and the firm will simply go ahead and log as long as net profits from doing so are positive (v-c>0). Hence, the firm will always win the conflict and establish *de facto* rights over the resource. Now consider the case where $s<b$. Then the benefits from protecting the forest for just one period (b) already outweigh the costs of blockading in that period. Thus, the community will always fight, and the firm, knowing this, will withdraw.

For intermediate values of b ($sr^C < b < s$), the boundary condition, represented by the line $v=\Omega$ in figure 3.1, is derived by computing for each player the maximum time that he can stay in conflict and still receive a non-negative payoff and equating these maximum times (see Burton, 2004). If $v>\Omega$, the firm is able to stay in conflict longer than the community and thus the firm obtains *de facto* property rights (or a situation of open access results). If, $v<\Omega$, then the community is able to stay in conflict longer than the firm, and thus establishes effective property rights over the resource.

From this analysis the probability that the community is able to establish *de facto* property rights (*PR*) can be written as:

$$PR = g(v, b, r^C, r^F, c, s). \tag{1}$$

As argued in Engel, López and Palmer (2006), inspection of the boundary condition ($v=\Omega$) indicates that:

$$\frac{\partial PR}{\partial v} \le 0, \frac{\partial PR}{\partial b} \ge 0, \frac{\partial PR}{\partial r^C} \le 0, \frac{\partial PR}{\partial r^F} \ge 0, \frac{\partial PR}{\partial c} \ge 0, \frac{\partial PR}{\partial s} \le 0. \tag{2}$$

Thus, the community is more likely to obtain *de facto* rights over the resource if the profitability of logging, blockading costs, and the community's discount rate are low, and/or if the community's valuation of the standing forest, logging costs, and the firm's discount rate are high. The intuition is as follows. An increase in an actor's benefits from winning (v for the firm, and b for the community) allows this actor to stay in conflict longer, and thereby raises the likelihood that he is able to win the conflict. An increase in fighting costs (c for the firm, and s for the community) has the opposite effect. Similarly, an increase in a player's interest rate induces him to value the immediate fighting costs more than the long-run benefits from winning and thereby reduces the maximum length of time this actor would stay in conflict and thus his chances of winning.

3.2.3 Community-company Bargaining Over a Logging Agreement

Engel, López, and Palmer (ibid) also present a bargaining model to capture community-firm negotiations over a logging contract, using the Nash bargaining solution. This is used rather than a non-cooperative game representation for two reasons. First, in the Indonesian setting, the local government often takes on the role of a mediator between firms and communities. Second, it has been shown that the Nash bargaining solution also represents the solution to a non-cooperative game of alternating offers where there is some probability that rejection of an offer will lead to a breakdown of the bargaining process (Gintis, 2000). It is assumed that negotiations are costless.

Engel and López (2004) present a more general model that allows for asymmetric bargaining power. This model is applied here, although simplified in the following ways. First, any possible differences in the logging area between unilateral logging and a negotiated agreement are abstracted from in this application. Second, it is assumed here that the community's reservation utility (d^C) is fixed, while Engel and López (ibid) allow for endogenous reservation utilities as a consequence of third-party interventions that are sensitive to the extent of logging. Both simplifications arise from the fact that the focus here is on negotiations over payments only. Negotiations over logging area are not considered. Engel and López (ibid) also explicitly decompose the community's valuation of the standing forest into the true value and the proportion of this value considered by the community. This is not necessary for the purpose of the present study. With these simplifications, the negotiated payment to the community (π^C) and to the firm (π^F) can be written as a standard asymmetric Nash bargaining solution to:

$$\max_{\pi^C,\pi^F} \left[\pi^C - d^C\right]^{\tau} \left[\pi^F - d^F\right]^{1-\tau} \quad s.t. \quad \pi^C + \pi^F = v - c, \tag{3}$$

where d^C and d^F are, respectively, the community's and firm's reservation utilities, and τ is the community's bargaining power vis-á-vis the firm ($0 \le \tau \le 1$). Muthoo (1999) shows that in the alternating offers game:

$$\tau \equiv \frac{r^F}{r^C + r^F}.$$

Thus, τ is generally increasing (decreasing) in the firm's (community's) discount rate, but may also depend on other factors associated with higher bargaining power (p). Hence:

$$\tau \equiv f\left(r^C, r^F, p\right) \text{ with } \frac{\partial \tau}{\partial r^C} < 0,\ \frac{\partial \tau}{\partial r^F} > 0,\ \frac{\partial \tau}{\partial p} > 0. \tag{4}$$

Solving (3) yields:

$$\pi^C = d^C + \tau\left(v - c - d^C - d^F\right)^{29} \tag{5}$$

The total 'cake' to be divided in negotiations are the net profits from logging (v - c). Thus, the asymmetric Nash bargaining solution implies that each player obtains his reservation utility, and the remaining surplus is divided in proportion to bargaining power. The reservation utilities of the players are determined by the outcome of the latent conflict over property rights. If the firm wins the conflict, its reservation utility is given by its logging profits (v - c), while the community's reservation utility is zero. If, by contrast, the community wins the attrition war, then its reservation utility is given by the present value of the standing forest:

[29] An implicit assumption here is that $v \ge d^C + d^F$. Moreover, the formulation by Engel and López (2004) assumes that both players are risk-neutral.

$$\left(\int_{t=0}^{\infty} be^{-rt} dt = \frac{b}{r^C} \right),$$

while the firm's reservation utility is given by its profits from using its capital elsewhere ($\bar{d}^F$).[30] In summary,

If $v>\Omega$, then $d^C = 0$ and $d^F = v - c$.

If $v<\Omega$, then $d^C = \frac{b}{r^C}$ and $d^F = \bar{d}^F$.

Substituting in (5) it then follows that:

$$\pi^C = \begin{cases} 0 \ \text{if} \ v > \Omega, \\ \frac{b}{r^C} + \tau\left(v - c - \frac{b}{r^C} - \bar{d}^F \right) \text{if} \ v < \Omega. \end{cases} \quad (5')$$

The model thus implies that the community's payoffs from negotiations depend crucially on its ability to establish *de facto* property rights (*PR*). From (5'):

$$\pi^C = h(v, c, b, \tau, \bar{d}^F, PR)$$

with $\frac{\partial \pi^C}{\partial v} \geq 0, \frac{\partial \pi^C}{\partial c} \leq 0, \frac{\partial \pi^C}{\partial b} \geq 0, \frac{\partial \pi^C}{\partial \tau} \geq 0, \frac{\partial \pi^C}{\partial \bar{d}^F} \leq 0, \frac{\partial \pi^C}{\partial PR} \geq 0.$ (6)

Using the expressions for *PR* and τ derived earlier (given in (1), (2), and (4)) in (6), the following reduced-form relationship between π^C and the model parameters is obtained:

$$\pi^C = \tilde{h}(v, c, b, r_F, r_C, \bar{d}^F, s, p) \text{ with}$$

$$\frac{\partial \pi^C}{\partial v} \gtrless 0, \frac{\partial \pi^C}{\partial c} \gtrless 0, \frac{\partial \pi^C}{\partial b} \geq 0, \frac{\partial \pi^C}{\partial r^F} \geq 0, \frac{\partial \pi^C}{\partial r^C} \leq 0, \frac{\partial \pi^C}{\partial \bar{d}^F} \leq 0, \frac{\partial \pi^C}{\partial s} \leq 0, \frac{\partial \pi^C}{\partial p} \geq 0. \quad (7)$$

The implications of these first order conditions are discussed in detail below.

3.2.4 Theoretical Hypotheses

3.2.4.1 The Effects of Parameters on Individual Stages

Local logging entrepreneurs, unlike outside logging companies before decentralization, were under social pressure to contribute locally (McCarthy, 2001). Hence, they could never really get away with providing no benefits whatsoever and always provided some minimal level of benefits to communities. This

[30] It is assumed here that $v>\bar{d}^F$ and $v \geq \frac{b}{r^C} + \bar{d}^F$.

minimal payment may be motivated by the idea that firms want to 'keep their face', and build and maintain political capital with the local government officials who issue the logging permits. Thus, it is determined by the relation between the company and the local government, and not by community-firm negotiations. But payments beyond this minimal level are the outcome of a two-stage game as described above, in which actual payments are negotiated. Moreover, where property rights are not formally defined and/or government enforcement is weak, actual payments beyond this minimal level will depend on the community's ability to self-enforce its property rights over the forest *vis-à-vis* the firm.

As described, the *de facto* property rights over the forest can be considered as the outcome of a war of attrition between the community and the firm. In this case, c can be interpreted as the fixed cost to the firm of a unilateral logging attempt by the firm including the minimal payment made to the community. Thus, when considering the actual payments made by the firm, if the community is not able to establish *de facto* property rights ($v>\Omega$ in figure 3.1) a situation of open access results and the firm would not make any actual payments beyond this minimal level. By contrast, if $v<\Omega$, the community can credibly threaten to stop any unilateral logging attempt. In this case, the firm has the option to either make the agreed payment or to stop logging altogether. The community's actual payoffs from logging can be represented by the Nash bargaining solution in equation 5'.

These results yield a two-stage process that can be hypothesized as follows:

Hypothesis set 1 (Probability of obtaining a payment above the minimal level):

(i) The community effectively shares in the benefits from logging by obtaining a level of actual payment above the minimal level only if it is able to self-enforce its property rights over the forest.

(ii) The probability that this is the case is increasing in fixed logging costs (c), the standing value of the forest considered by the community (b), and the firm's discount rate (r^F), and decreasing in the profitability of timber (v), blockading costs (s), and the community's discount rate (r^C).

Hypothesis set 2 (Size of payment received):

(i) If the community is able to self-enforce property rights over the forest, its payoffs per hectare of forest logged are given by the Nash bargaining solution in equation (5).

(ii) As described in equation (6), these payoffs are likely to be increasing in the profitability of logging (v), the standing forest value considered by the community (b), the firm's discount rate (r^F), and the community's bargaining power *vis-á-vis* the firm (τ), and decreasing in the value of the firm's reservation utility (d_F) and the community's discount rate (r^C).

This interpretation of the model implies three important discontinuities in the effects of the model parameters. First, the community's cost of blockading (e.g., its ability for collective action) affects only the likelihood that the community is able to self-enforce property rights and thus receive a payment above the minimal level; it does not affect the actual payment negotiated. By contrast, the community's bargaining power only affects actual payments, but not the community's ability to enforce property rights. Finally, an increase in the profitability of logging decreases the probability that the community can self-enforce property rights over the forest, because it increases the incentives for the firm to fight and log unilaterally. For those communities that can win such a fight (i.e., can enforce property rights) an increase in the profitability of logging, however, increases the community's payoffs, because it increases the size of the cake to be divided.

In order to test hypothesis set 1, the logistic probability model or logit model for a binary outcome is used in chapter 12[31]. The reduced form of the model is:

$$\text{Prob}(y_i) = \frac{e^{\beta' x_i}}{1+e^{\beta' x_i}}, \qquad i = 1, 2, \ldots N, \tag{8}$$

where N = number of communities in the sample and y_i denotes payoff to community i. All payoff values at or above a specified minimum level will be designated the value 1 ($y_i = 1$ if $y_i^* \geq 0$. y_i^* is unobserved) and any payoff values below this level, will be designated the value 0 ($y_i = 0$ if $y_i^* < 0$).

In chapter 12, hypothesis set 2 is tested using ordinary least squares (OLS) regression techniques, and only for those payoff values at or over the minimum level. The model follows the usual linear form:

$$y_i = \beta_0 + \beta_1 x_{1i} + \beta_2 x_{2i} + \varepsilon_i, \qquad i = 1, 2, \ldots N, \tag{9}$$

where N = number of communities in the sample, y_i denotes payoff to community i, while X_{1i}, X_{1i} denote the independent variables (proxies for profitability of logging, standing value of the forest to the community etc). Given the large number of potential proxies for the variables in the model, a forward stepwise regression method is utilized. Heteroskedasticity is anticipated in both the logit and OLS models, which has to be corrected for in the analyses.

3.2.4.2 The Total Effects of Parameters

The hypotheses presented below combine the direct effects that parameters have on community payoffs through bargaining (equation (5')) and the indirect effect

---

[31] See Greene (2000) for more on the logit and other qualitative response (QR) models (p. 811).

that parameters have on payoffs by influencing the probability that the community will win a potential conflict (equations (1) and (2)). The total effects of parameters are given in equation (7). Some factors have both a direct and an indirect effect that move in the same direction. For example, an increase in the valuation of the standing forest by the community raises community payoffs directly and also increases the community's ability to win a potential conflict. For other factors, the direct and indirect effects move in different directions. For example, an increase in the profitability of logging (v) has a positive direct effect on community payoffs as increased profits imply a larger 'cake' to be divided. The indirect effect of an increase in logging profits is, however, negative. This is because a greater profitability of logging raises the firm's willingness and ability to fight for *de facto* property rights, thus reducing the community's chances of winning a conflict. Thus, the total effect of an increase in logging profits on community payoffs is *ex ante* ambiguous. Which effect dominates depends on initial conditions and is essentially an empirical issue. Finally, some factors have only one effect. For example, blockade costs enter only indirectly, as they affect conflict outcomes and thereby reservation utilities, but they have no direct effect on logging.

In summary, the model yields the following hypotheses:

Hypothesis set 3 (combined effect):

(i) Community payoffs, π^C, are likely to be increasing in the community's valuation of the standing forest (b); the firm's discount rate (r^F) and other factors leading to an increased bargaining power of the community (τ).

(ii) π^C are likely to be decreasing in the community's discount rate (r^C); the firm's profits from using its capital in the next-best activity ($\bar{d}^F$); and the community's blockading costs (s).

A change in timber profits, v, on the other hand has an indeterminate effect on π^C due to the counteracting effects explained above. The same holds true for an increase in fixed logging costs (c).

Similar to hypothesis set 2, ordinary least squares (OLS) regression techniques are utilized for the testing of hypothesis set 3 in chapter 12 (see equation 9). The main difference is that the entire sample is included in the analysis and not just a subset of the sample. Again, given the potential number of proxies for the variables in the model, a forward stepwise regression method is utilised, and potential heteroskedasticity has to be corrected for in the analysis.

In chapter 12, the model is operationalized by defining proxies for each theoretically relevant factor and by testing econometrically the hypothesised effects of the model parameters for both the individual stages (hypothesis sets 1 and 2) and the total effects of parameters on outcomes (hypothesis set 3).

4 Research Methods and Questionnaires

The apparent similarities in findings from case studies of IPPK/HPHH deals led to the development of community level and household level questionnaires to allow for a direct comparison of contractual and actual outcomes among a relatively large sample of communities in East Kalimantan. Between March and June 2003, a preliminary field trip to East Kalimantan was undertaken to scout out prospective survey communities, initiate the development of the questionnaires and build a research team. This chapter describes the development of the questionnaires, field site selection and the conduct of the fieldwork.

4.1 Development of Questionnaires

The design, structure and nature of the surveys and the questions within were motivated in two ways. First, during the preliminary field trip to East Kalimantan, a two week long stay with some communities in Bulungan was undertaken. Information relating to IPPK agreements was gathered via informal interviews with community leaders and other community members resulting in a case study (see Palmer, 2004). The results formed the basis of the two questionnaires. The information gathered was cross-checked and substantiated with interviews with local NGOs active in the area and Forest Office personnel in Tanjung Selor, Bulungan. This study attempted to crystallize the main issues and the means by which information and data could be efficiently obtained for the purpose of a larger-scale comparative survey. The direct comparison of cases is justified by the common use of production-based fees (Rp. per m^3) in logging agreements and in calculations in the payments for communities. Second, the model adapted from those developed by Engel, López and Palmer (2006) and Engel and López (2004) led to a set of hypotheses as described in chapter 3. To test these hypotheses, empirical proxies were required that would fit the various parameters from the model, in addition to a large enough dataset to permit regression analysis. Thus, the questions in the surveys were designed to capture the necessary information for these empirical proxies.

Two questionnaires were developed between June and September 2003. One was designed for community leaders and people who had directly involved in negotiations with logging companies, and the other for households in the wider community who were supposed to benefit from these negotiations. The process of negotiations with logging companies usually only involved a sub-set of the community, typically the community elite and leaders (see chapter 7). The unit of measurement for this survey was a community, defined as a group of people that had negotiated a single agreement or contract for the benefit of all members of that group. Members in the group were usually connected by family and ethnicity, with a historical, customary (or *adat*) forest claim. Thus, within the same village (known as a *desa* or *kampung*, an administrative unit formulated by central government and modelled on the Javanese system), there were sometimes multiple communities present. They were typically grouped together for administrative convenience despite sometimes having separate ethnic identities and claiming separate forest territories. This therefore explains why there needed to be a clear distinction made between the village and the community on the ground, although in this study hereafter the terms 'village' and 'community' will be used interchangeably. Where the term 'village' is used in the governmental context, it will be prefixed by 'administrative'.

The community level questionnaire was constructed for the focus group format because discussions relating to community matters such as IPPK/HPHH agreements typically involved the men, and only very occasionally, the women, responsible for decisions on behalf of all in the community. Typically, agreements would normally involve the village head (*kepala desa*) and the customary law head (*kepala adat*). With this in mind, a large number of short questions were designed and grouped in six sections (see table 4.1). Almost all the questions followed a 'closed format' rather than being open-ended. Thus, while there were always spaces available for comments, questions were either framed around 'yes', 'no' or 'don't know' responses or a list of possible responses accompanied with tick boxes. This format deliberately limited the opportunities for interpretation of the questions in the field, which was necessary to enable a direct comparison of responses among sampled communities.

Overall, the sections in the community survey were designed to follow an approximate, chronological sequence of events, with the first focusing on basic community characteristics. In a focus group situation, the discussion usually began with section B on pre-decentralization experiences and moved from there. The first section was reserved for a small, separate meeting with the village secretary (*sekretaris desa*) who was normally the one responsible for the collection of basic community statistics (population, ethnicity, employment etc). In some cases, these statistics were cross-checked with data collected from other surveys, such as those undertaken by CIFOR. Of course, there was nothing to stop a discussion moving away from a predetermined sequence of questioning and the clear structure of the questionnaire allowed for unpredictable changes to the thread of the discussion. It was important to be prepared for this given the group dynamic in each community.

As the discussion moved along the thread of each section, the community questionnaire was designed to capture precise answers in a relatively logical way. Beginning with experiences of dealing with logging companies before decentralization, questions turned to experiences of community members of working for logging companies in the past. At the end of this section, questions were asked regarding knowledge of the IPPK/HPHH system before an approach was made to open negotiations on a possible contract. Section C concentrated on the approach made and the actual negotiations; who spoke to whom, when and where, and what was discussed. Section D focused on the actual agreement made, with information to be gathered on the terms of the trade-off agreed by the community and the company. The following section, E, was broader in scope, asking for information on the outcomes from these negotiations, although the type of information requested was similar to that in section D. Questions were asked on area logged, benefits received, the distribution of these, and so on to problems within and with other communities as a consequence of the IPPK/HPHH system. The second half of section E focused almost entirely on company non-compliance and the community response to this, and also included questions on any perceived environmental impacts from these operations. It was anticipated that each group discussion would last up to two hours and that the group format would lead to accurate and reliable answers, particularly with regards to the contracts and outcomes. Box 4.1 illustrates the motivations for all the questions in the community survey in forming proxies for the model parameters, in addition to the kind of information requested (see chapter 12).

Table 4.1 Structure of the Questionnaires

Section	Household questionnaire	Community questionnaire
A	Basic household characteristics	Basic village characteristics
B	Index of assets	Community experiences of logging before decentralization
C	Household income and livelihood	Experiences about the process of negotiations before logging agreement made
D	Household experiences of logging before decentralization	Information about the first agreement made with the company
E	Household participation in and knowledge of logging agreement	Information about the outcome of the first agreement
F	Information on benefits from agreement	Other information
G	Information on any conflicts between the community and the company	
H	Attitudes of household towards customary forest	
I	Alternative scenarios	

Box 4.1 Motivation for Questions in the Community Survey

A1.	Community's location and degree of isolation relative to nearest town/city (ease of access for firms, outside influence), community forest area claim, distance to nearest market (opportunity costs), length of time community had resided in current location (strength of property rights)
A2.	Information on neighbours, border agreements made and when (land claims, strength of property rights)
A3.	Community population size, ethnic groups, patterns of in-migrations (collective action)
A4.	Employment by category and wages (degree of dependence on forest for livelihoods)
A5.	Land and forest area and utilization (forest management and property rights)
A6.	Attempts to register land titles (strength of property rights)
A7.	Level of community participation in forest management decisions (social capital, collective action)
A8.	Community participation in general (collective action)
A9.	Infrastructure before IPPK/HPHH began (indicator of local material development/poverty)
A10.	General outside assistance to village for development, agriculture etc.
B1.	Pre-decentralization logging experiences: compensation paid, information on conflicts, experience of negotiations, perceptions of environmental impacts (bargaining power, valuation of local forest)
B2.	Previous knowledge of IPPK/HPHH systems (bargaining power)
B3.	Previous experience with working for logging companies (bargaining power)
C1.	First approach by IPPK/HPHH, who was involved, where did meetings take place, what was discussed, community participation
C2.	Points of discussion during negotiations
C3.	Acceptance of first offers on terms in agreement
C4.	Information on involvement of third parties during negotiations and approaches by other firms
D1.	Information about first agreement: length, type of agreement, parties involved, how agreement was made
D2.	Contents of first agreement: benefits promised, logging rules, area agreed
E1.	Outcome of first agreement: area, length of time, money paid, distribution of benefits, problems in village, system of monitoring company's activities, overlaps between IPPK/HPHH and HPH (strength of property rights) and conflicts with other third parties, e.g. other communities
E2.	Non-compliance by company: nature of non-compliance, actions taken by community and how resolved, role of government, details on any new agreement made, location of new discussions, outside assistance. Future perceptions
E3.	Perceptions of environmental impacts of IPPK/HPHH
F1.	Awareness of alternatives (uncertainties and strength of property rights claim)
F2.	Attitude of village head towards forest claim

The household survey on the other hand was far shorter and designed more for one-to-one question and answer sessions, lasting up to an hour. It was not designed as a household income survey (see table 4.1). Similar to the community survey, the questions in the household survey followed a closed format. While the questionnaire was primarily designed for the head of the household (known as a *kepala keluarga* or *KK*, who was usually male), many of the questions were similar to those asked to the village leaders. For example, section D in the household survey on experiences of dealing with logging companies before decentralization, has similar questions to those contained within section C in the community survey. In section E, households were asked about who met whom in the meetings between the community and the company prior to making an agreement, and whether they participated in any of these. This information from a sample of households in any given community was to be compared with the answers given for similar questions asked in section C of the community survey.

Thus, while the community survey was the more important and comprehensive of the two surveys due to the use of communities as the principal unit of measurement, the household survey allowed for the construction of more precise indices of community characteristics, particularly where data were not available from the secretary. The household data was also to be used to construct community indices, for example, on household participation and awareness of what was happening within their respective communities. Perhaps more importantly, household responses could be used to substantiate the information collected from the village leaders, and check it for inaccuracies or biases. Where information did not corroborate, particularly in cases where community leaders may have been less cooperative or less knowledgeable, the household survey answers became more important in attempting to obtain the most accurate and reliable picture possible of the IPPK/HPHH situation in the community. Box 4.2 gives the structure and breakdown of the questions in the household survey.

Box 4.2 Motivations for Questions in the Household Survey

A.	Details of family, ethnicity, education, access to healthcare, employment, length of residence in village, participation in village decision making and organizations
B.	Index of prosperity: description of house and assets, and impacts from IPPK/HPHH
C.	Household livelihoods, farming, experience of employment with logging companies, forest product collection, impacts of IPPK/HPHH
D.	Household experience of logging operations before decentralization: compensation, conflicts, environmental damage
E.	Household participation in IPPK/HPHH negotiations, knowledge of parties, decision making, prior expectations of benefits
F.	Household perceptions of company promises to community, promises to household, identity of who received benefits first, what was received by household and future expectations, intra-community problems
G.	Company non-compliance and problems, monitoring and enforcement, conflicts (frequency and intensity) and household participation in these, identity of leaders and other participants
H.	Household perception of and attitudes towards community forest
I.	Awareness of alternatives to IPPK/HPHH

4.2 Community and Household Selection

The guideline in site selection was that all communities were to have been involved in small-scale, mechanised logging agreements, i.e. involving negotiations between representatives of the communities and the timber buyers and capital owners. While unilateral logging by firms without community consent or agreement still took place on land claimed by communities, this was not widespread in East Kalimantan. Many communities with credible claims were able to secure agreements with companies even if these were only very informal, oral agreements. Furthermore, all of these agreements were to be legitimised by the issuance of permits by district heads. For direct comparison, only those agreements that took place prior to companies entering community territory and initiating logging activities were to be considered[32]. For communities that had negotiated multiple agreements, only the first one was to be surveyed.

---

[32] It was very difficult for firms to log unilaterally on land that was claimed by communities without making any kind of agreement with them in the immediate post-Suharto era. There were cases in which firms logged without securing an agreement first, although protests by community members would ensure that an agreement was made *ex-post* even if no permit was applied for. These kinds of agreements were specifically excluded from this study. On close investigation, however, two of the surveyed communities had this kind of arrangement. They were excluded from further analysis (see chapter 5).

Communities were selected on the basis of information and research that had already been undertaken, e.g. from case studies, lists of IPPK and HPHH permit lists from district Forestry Offices and numerous discussions with local researchers and NGO workers, in addition to CIFOR staff in East Kalimantan and Java. All permits issued in Malinau and Bulungan were IPPK, while those in Kutai Barat tended to be HPHH. As discussed in chapter 2, there were no real differences between these two types of permit. Both were used as means of operationalising small-scale, capitalised logging concessions and in providing use rights to local communities. Also, both types of permit allowed for bargaining between communities and companies in the absence of guidelines from local governments or any other third party. Thus, they are directly comparable and any differences between them are described in the text in the following chapters.

Approximately 22 communities were selected from each of the chosen administrative districts in East Kalimantan, making a total of 66 communities. For each district, communities were selected on a sub-district basis to give a broad geographical spread and according to the constraints of the research budget and time available in the field. While an effort was made to sample these communities as 'randomly' as possible, the circumstances in the field meant that there was an element of selectivity bias in the sample. From the beginning it was acknowledged that it probably would not be possible to sample all 66 communities, although at least 50-55 were required to enable the use of econometric analytical techniques.

Even with the guidelines and resources available, very little was really known about the situation on the ground until the fieldwork commenced in September 2003, particularly outside previous case study research areas. Data collected from district governments such as permit lists were very unreliable and not always up to date. For example, in Malinau by late 2003, the total number of IPPK agreements according to the district government stood at 46 (Limberg, 2004). Of these, some communities had multiple agreements, while other agreements were never operationalized in the first place. Also, a number of agreements in Malinau were in effect shared between communities and hence were not separate, independently negotiated contracts as indicated on the permit list. These kinds of details were not necessarily available to the research team prior to the commencement of the fieldwork. To complicate matters in Malinau, other fieldwork had been carried out in some communities for other projects in 2003, giving rise to possible 'respondent fatigue', so a decision was taken to avoid some of these places as well. In the end, a very high proportion of all active IPPK agreements were sampled in Malinau. Similar problems were encountered in Bulungan and Kutai Barat as well. For example, one report suggested that only 30 percent of the 800 HPHH permits issued in Kutai Barat were operational in 2003 (Kompas, 2003). Thus given the lack of details, it was important to be flexible and prepared to move on to the nearest, suitable communities in which permits had been operationalized. Nevertheless, similar to Malinau, a high proportion of all active agreements were sampled in both Bulungan and Kutai Barat.

In most communities a standard 10 households were randomly sampled, with the exception of very small communities composed of between ten and 30 households, and very large communities with hundreds of households. In the former, between five and seven households were usually sampled and in communities containing between 200 and 500 households, 15 households were interviewed. In communities containing over 500 households, 20 of these were interviewed. In the larger communities, people tended to be divided along administrative lines into sub-community units (known as *Rukun Tetangga* or *RT*) and these were not necessarily drawn along ethnic lines. For sampling purposes, the selection of households was usually split along the lines of ethnic groupings and income levels within the community. So, in a community of say 100 households comprising two distinct ethnic groupings, these would be sampled according to the approximate proportions of each grouping. Once this had been calculated, where each grouping lived in the community was identified and an attempt was made to further sub-divide these by income. Since income data were only very rarely available, 'house quality' was used to identify households on the basis of wealth (see section B in box 4.2). Once a rough breakdown was available, households were randomly selected by assigning a number to each house and literally 'pulling numbers out of a hat'. Households were then approached and if people were unavailable at that time, researchers moved to the next house or attempted to return at a later time or date. Prior to starting the interview, households were asked if they had lived in the community prior to negotiations for the first IPPK/HPPH concession. If not, the researchers were instructed to move to the next household on the list.

In both the selection of communities and households, there were possible sample selection biases. Correcting for these required knowledge about the numbers of communities with logging agreements in each district and sub-district, and the numbers of households in every community. Much of this information was not always available, accurate or reliable in advance of the fieldwork. As mentioned before many of the communities had not been surveyed before and government data was for the most part, questionable. Even after the fieldwork had been completed some of the necessary information was still unavailable. Therefore, the descriptive statistics contained in this study, in particular those presented in Part II, may be biased and should be approached with a degree of caution.

4.3 Fieldwork

During the course of 2003, and through visits, interviews and meetings in Java and East Kalimantan, a research team[33] was put together and placed on three to four month fieldwork contracts in September 2003. The team met together for the first time at a training workshop held in Tanjung Selor, Bulungan in end-September 2003. It lasted four days and included a questionnaire 'field-test' in the community of Jelerai Selor near Tanjung Selor. In addition to ironing out problems with the questionnaires and explaining the purpose of the questions from the point of view of the research objectives, there were opportunities to discuss travel plans and develop a team spirit. Separate meetings were held with individuals and pairs on specificities for each research area, including plans on which communities to visit given the information that was available at that time, and budgeting issues.

For each community, the process of engaging people in the surveys would begin with the researchers making an approach to the village head and/or other community leaders. Here, there would be an opportunity to explain the purpose of the visit and the research to the leaders. Before permission was sought and any further arrangements made, the research team was issued the following instructions on the selection of communities for the survey:

- Has the community made an IPPK/HPHH agreement with a company?
- Has the company already started and completed logging operations, i.e. was the first agreement finished?
- Did the community have a shared agreement with another community?
- Did these communities receive separate benefits from the company?
- Did these communities have separate, claimed forest areas?

For shared agreements, the researchers were instructed to attempt to visit all communities sharing a single agreement. The responses to (3), (4) and (5) were used in forming the parameters for each community involved in these agreements. Most importantly, conditions (1) and (2) had to be satisfied before going any further, otherwise the researchers were instructed to move to the next community on their respective lists. If satisfied, it was of utmost importance to be properly welcomed into the community before making an attempt at arranging a meeting for the community survey and seeking permission for undertaking

[33] The team comprised six Indonesians, five of whom originated from East Kalimantan, and one from Java. Nopilus and Doni were employed by the NGO SHK Kaltim and had had previous experience of working in Kutai Barat. Yan Ngau and Samsu worked for another NGO, Pionir Bulungan, again with experience of working in communities in that particular district. Dodi Hernawan was employed by the NGO Biosfer-Manusia (BIOMA) and although based in Samarinda, had previous fieldwork experience in Malinau. He was joined by Rini Kusumawati, a Javanese student. Please note that many Indonesian people only use a single name.

household interviews. Additionally, information was to be obtained on the approximate size of the community, its ethnic composition and relative income levels. Where available, a list of households was to be obtained. These were used to begin the process of randomly sampling the households. It was envisaged that the community and household surveys would take an average of three to four days per community. While this was not a lot of time, the researchers were encouraged to talk to people in the community around the formal survey format to gather extra, qualitative information on the IPPK/HPHH situation in each community.

From September 2003 until January 2004, the team visited a total of 65 communities spread out among three districts, with two person teams in each district (see chapter 5 for full list of communities visited during this period). The three pairs of researchers worked in their respective districts simultaneously and had no contact with any of the other pairs until the fieldwork had been completed. I planned my time so that I could spend approximately a third of it in each research area, to help guide the other researchers and get a real understanding for what was going on in all the areas under study. While moving from district to district, I was able to share the problems from the surveys and new ideas on how to approach the interviews with each pairing. A total of 65 community-level and 687 household interviews were conducted in the districts of Malinau, Bulungan and Kutai Barat, from September 2003 until January 2004. In January 2004, an end of fieldwork workshop was held at the offices of the NGO SHK Kaltim in Samarinda. This gave the team an opportunity to discuss problems in the field and exchange observations and findings with one another. In light of the descriptive and empirical results presented in Parts II and III, some of these problems are discussed in chapter 13.

5 Community and Household Characteristics

65 communities were sampled of which three were dropped due to incompatibility with the remainder[34]. These 62 communities are listed in table 5.1, along with the years that they were established in their current locations.

Table 5.1 Communities Sampled in East Kalimantan and Year of Current Settlement Foundation, September 2003-January 2004

Malinau		Bulungan		Kutai Barat	
Sub-district Malinau West		*Sub-district Tanjung Palas*		*Sub-district Melak*	
Batu Lidung	<1963	Antutan	1700s	Sakaq lotoq	<1952
Sesua	1948	Mara Satu	1700s	*Sub-district Muara Pahu*	
Punan Bengalun	1941	Pejalin	1800s	Kendesiq	<1950s
Sub-district Malinau South		Long Beluah	1900s	Abit	<1963
Setarap	1920s	Mara Hilir	1900s	*Sub-district Damai*	
Punan Setarap	1940s	Sajau Metun	1964	Temula	<1952
Long Adiu	1880	*Sub-district Peso Hilir*		*Sub-district Bongan*	
Punan Adiu	1900s	Long Telenjau	1800s	Muara Siram	1926
Langap	1914	Long Bang	1963	*Sub-district Long Bagun*	
Nunuk Tanah Kibang	<1971	Long Peso	1890	Ujoh Bilang (Pangyung Urip Batu Ayau)	1936
Tanjung Nanga	1974	Long Tungu	1965	Ujoh Bilang (Bitt Unyang)	1936
Sub-district Mentarang		Long Lasan	<1940s	Long Bagun Ulu	1900s
Long Sulit	1982*	*Sub-district Tanjung Selor*		Batoq kelo	1983
Long Simau	1980*	Jelarai Selor	1985	Batu Majang	1983
Paking Lama	1940s	Gunung Seriang	<1941	Long Melaham	1970
Sub-district Malinau North		*Sub-district Sesayap*		Mamahak Besar (Luuang Karimam)	1901
Malinau Seberang	1920s	Sebiday	1972	Mamahak Besar (Sungai Mas)	1901
Tanjung Lima	1900s	Sedulun	1969	Mamahak Besar (Mayang Putri Prima)	1901
Tajan	1960s	Kujau	1950	*Sub-district Long Iram*	
Sebatiung	1900s	Limbu Sedulun	1960	Long Daliq	1950
Kaliamok	1968			Keliwai	1800

[34] Both Semolon in Malinau and Naha Aya in Bulungan negotiated IPPK benefits after the company had already entered and begun operating on its territory. Kenyanyan in Kutai Barat had a HPHH permit for a non-mechanised operation and sold the timber directly to buyers for varying prices.

Table 5.1 Communities Sampled in East Kalimantan and Year of Current Settlement Foundation, September 2003-January 2004 (continued)

Malinau		Bulungan		Kutai Barat	
Sub-district Malinau North		*Sub-district Sekatak*		*Sub-district Long Iram*	
Kaliamok	1968	Sekatak Bengara	1834	Keliwai	1800
Kelapis	<1977	Bekeliu	1969	Muara Mujan	1930
Salap	1927	Punan Dulau	1970	Ujoh Halang	1990
Selidung	1930s	Sekatak Buji	1800s	Kelubaq	1991
				Kelian Luar	1900

Note *These communities also have members located further upstream in previous settlements that were founded in the 1900s. '<' denotes that the community existed in some form prior to this date, although no more information was available at the time of the survey.

The sampled communities are grouped according to administrative district, mainly because this was the unit which issued the IPPK/HPHH permits in the first instance: 21 in Malinau, 21 in Bulungan and 20 in Kutai Barat. Communities were sampled from a total of 15 sub-districts and had lived in the same locality for an average of nearly 80 years, with the shortest period being 12 years and the longest, 300 years.

In the administrative village of Ujoh Bilang, there were five separate negotiating groups, comprising 50 percent of the village's total population. These people called themselves the 'original inhabitants' (*orang asli*) of the village, with the other 50 percent classified as 'newer arrivals'. Thus, in addition to 'original inhabitant' status, membership of these groups was normally based on family ties to the group and ethnicity. Each group claimed separate areas of forest and made separate agreements with different companies at around the same time, with separate and independent benefit schedules. These groups are defined as separate communities. Two of the five communities in Ujoh Bilang were sampled (see table 5.1). There was a similar situation in the administrative village of Mamahaq Besar, with three separate, negotiating groups. All three were sampled in this survey.

5.1 Nearest Town and Market

For many of the sampled communities, the nearest town was the same place as the nearest market (see figure 5.1). For a majority of communities in Kutai Barat, the nearest market was actually much closer than the nearest town, and in fact there was a market in five communities sampled in this district (so distance for these is '0'). In the remainder of the sample (42 communities), the only community with a market was Sekatak Buji in Bulungan. All communities in

Malinau were dependent on the market in the district capital, Malinau town, while all communities in Bulungan apart from those in the sub-district of Sekatak were dependent on the market in the district capital, Tanjung Selor.

Figure 5.1 Distance to Nearest Town and Marketplace for all Sampled Communities (km).

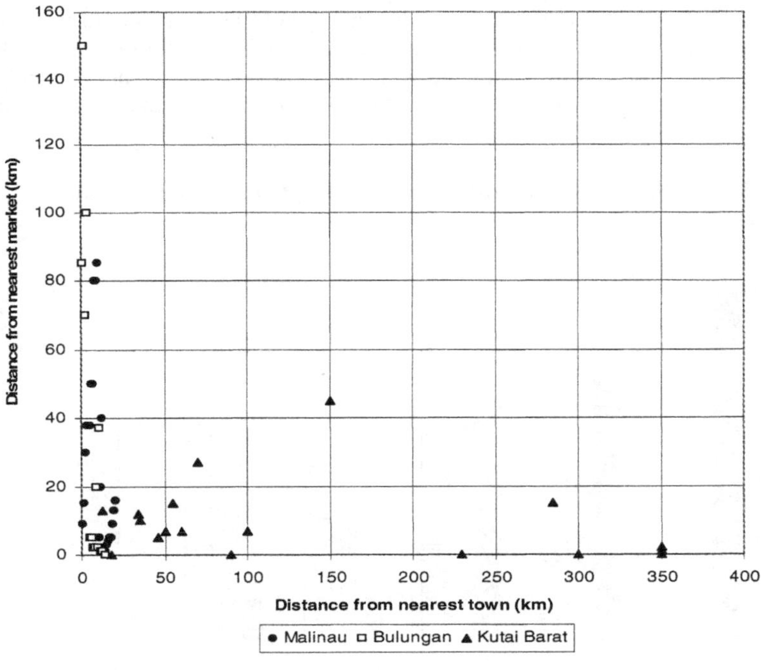

5.2 Population, Homogeneity and Ethnicity

The most common unit of measurement used for population in East Kalimantan was 'family heads' (*kepala kelarga* or KK). This is a good proxy for number of households. Overall, the average number of households in each community was 166, with large variation from the smallest (17 households) to the largest (950 households). Bulungan had the largest number of family heads per community on average (245), followed by Kutai Barat (163) and Malinau (91). Of 21 sampled communities in Malinau, eight had less than 50 family heads and only one had over 251. In Bulungan on the other hand, there were no communities smaller than 50 family heads and six with 251 or more.

Ethnic homogeneity in this study was defined as any ethnic grouping accounting for 80 percent or more of the total population. Typically but not always, communities could be distinguished by ethnicity. On average, 83 percent of households belonged to the dominant ethnic group in the community, varying from a low of 40 percent and high of 99 percent of households per community. Less than half of communities sampled in Bulungan (10 out of 21) did not have populations with a dominant grouping accounting for 80 percent or more of the total population. Both in Malinau and Kutai Barat, over 70 percent of sampled communities had dominant groupings with this proportion of the total population.

Figure 5.2 Numbers of family heads for sampled communities

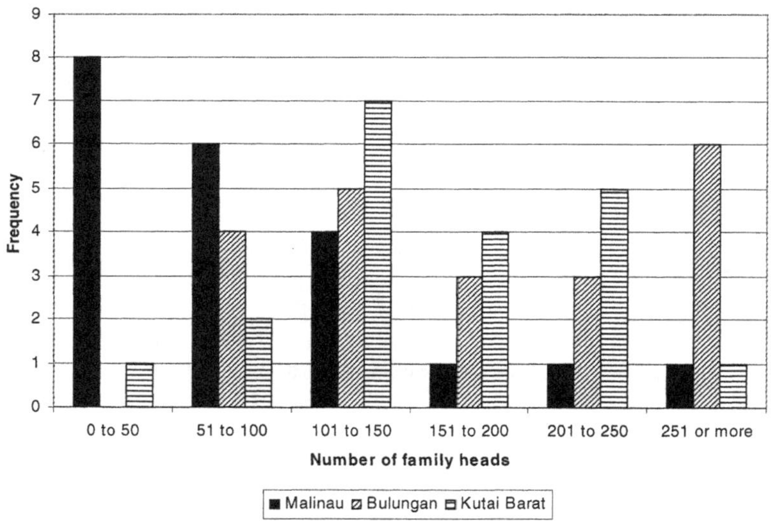

The definition of ethnic groups used in this study was dependent on how respondents in the household survey identified themselves in relation to other groups[35]. In Malinau, the dominant groups were the Brusu, Punan, Kenyah, Lun-

[35] It should be noted that some respondents that are grouped under some of the larger groupings such as the Kayan and Kenyah, defined themselves as a further sub-division of the ethnic headings used here.

daya (Putuk), Abai, Merap, Tagel and Tidung. Smaller groups in Malinau, with 20 to 79 percent of the population, tended to be similarly ethnically defined. In Bulungan, there were similar dominant groupings; Brusu, Kenyah, Tidung and Punan, along with large numbers of Kayan. The ethnicities of the smaller groups in Bulungan were similar to those of the larger groups too. The only dominant groups in Kutai Barat also found in the other two districts were the Kenyah and Tidung. Other dominant groups included the Bahau, Kutai, Tunjung, Bentian and Benuaq. Smaller groups in Kutai Barat included the Kayan (also found in Bulungan), Tunjung, Bahau and Bentian peoples.

There was an average of 1.3 separate ethnic groups in each community with a population that comprised 20 percent or more of the total population in the community. The most heterogeneous communities had a maximum of three separate ethnic groupings, each comprising 20 percent or more of the total community population. Almost all communities in the sample had much smaller proportions of households (five percent or less) of people belonging to different indigenous groupings, in addition to non-indigenous migrants from other parts of Indonesia such as Javanese and Bugis (from Sulawesi).

5.3 Neighbouring Communities and Forest Border Agreements

Figure 5.3 illustrates the numbers of neighbours each community in the sample had, i.e. communities with which each shared land and forest borders. As this figure shows, a majority of communities in Kutai Barat and Bulungan tended to have two neighbours only, while most in Malinau had three or more. The number of neighbouring communities bordering these forest claims varied from one to eight, with mean of almost three. The communities sampled in Bulungan and Kutai Barat were mainly to be found on major rivers such as the Mahakam or Kayan, while those in Malinau were to be found by a greater number of large rivers, such as the Malinau, Mentarang, Semendurut and Sesayap rivers. Therefore, while many of the communities sampled in Kutai Barat and Bulungan were situated along a single, major river, many communities in Malinau were clustered around river junctions. This may be reflected in the fact that of the 84 percent of all community-community relationships that were characterised by some kind of border agreement in Malinau, over 90 percent of these were made before IPPK/HPHH agreements were made. The other two districts were also characterised by having over 80 percent of neighbouring communities involved in some kind of border agreement.

Figure 5.3 Numbers of Neighbouring Communities for Each Sampled Community by District.

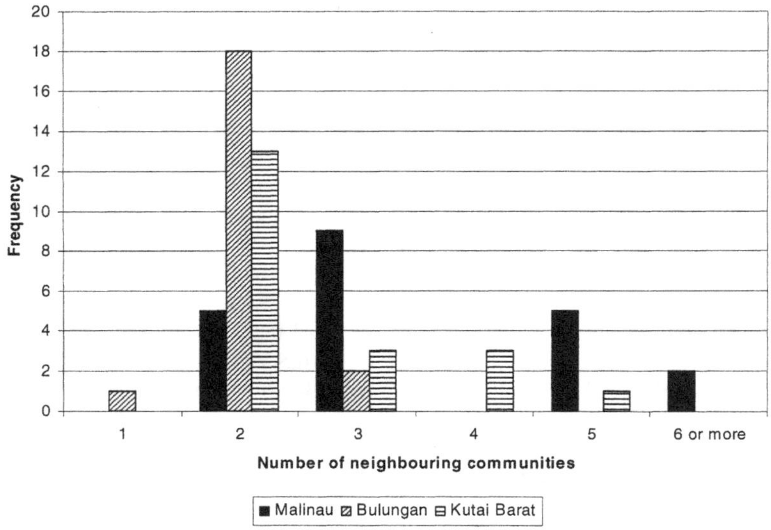

However by contrast, there were far fewer border agreements made before IPPK/HPHH agreements were negotiated. In Bulungan and Kutai Barat, only 39 and 60 percent of border agreements were made during or after the IPPK/HPHH agreement, respectively. Many if not all of these agreements were relatively informal and based on customary law. Moreover, some of these agreements covered 'village territory' and it was not always clear if this included forest land or not. Sometimes, village territory merely comprised farmland and gardens, with forest areas classified separately using terms such as customary forest (*hutan adat*) or village forest (*hutan desa*).

5.4 Employment and Livelihoods

A large majority of the sample had livelihoods based on subsistence farming in forest areas. Thus, on the whole the communities in the sample had very similar structures of employment and livelihoods. While incomes were not explicitly measured in this survey, they tended to be derived from farming and gardens and the sale and trade of agricultural surplus. Typically, most working household members had a plurality of productive activities depending on seasonality and opportunities, although subsistence agricultural activities predominated in the working lives of between 75 and 100 percent of households in each community, including women. These activities revolved around 1-2 ha of rice fields (*sawah*) per family with some gardens for fruit trees and vegetables. At the same time most people would keep some animals (chickens, goats, ducks, and in non-Muslim households, pigs) and fish in ponds in some places. Surplus would be sold in the village, or more likely the nearest marketplace. Farmers could earn up to USD 30-40 from surplus production per month.

In all communities, there were people working in the 'private sector', typically owning stalls or shops, working for logging operations or operating 'taxis' (boats, motorbikes), in around five percent of households. These kinds of activities were more pronounced in bigger communities and ones nearer to the towns, e.g. in Malinau Seberang and Jelarai Selor. In these communities, there were also more people (perhaps another 10 percent of households) working in commercial plantations, shrimp farms and in animal husbandry full-time. Even in these households however, people still engaged in rice farming and owned animals such as chickens and ducks.

Communities that also based the sub-district government or were located near these tended to have slightly more people employed as government officials. For example in Sekatak Buji, over 20 percent of households had at least one member working for the local government in some capacity. An average of over six percent of households in each community had at least one member working for the local government, for example as officials or teachers, with wide variation from 0 to over 40 percent of households. Incomes tended to be around USD 40 to 150 per month for government officials and private sector positions, depending on the position held and so on. Again, in many of these households there were household members that engaged in farming activities as well.

A very high proportion of households (mean 49 percent per community, ranging from 0 to 83 percent) gained experience of working in the logging industry prior to negotiations, either close to home or across the border in Malaysia. A lower proportion of households, 33 percent, had experience of working outside the confines of their own communities before IPPK/HPHH concessions. In some communities however, there were virtually no households with outside working experiences while in others, up to 80 percent of households had worked away from their respective communities.

The share of total household income derived from forest products varied from 5 to over 80 percent, with an average of 40 percent. Thus, the collection of forest products varied greatly from place to place, although it tended to be very seasonal. Some communities such as the Punan gathered a large variety of forest products and used them for domestic purposes, whereas others were more focused on the gathering of one or a few types of product solely for income generation, e.g. bird's nests and tree bark used in incense (known as *gaharu*).

5.5 Land Claims and Registration

On the basis of data collected at the community level, figure 5.4 illustrates land claim per community, in ha. Communities claimed on average over 300 ha of forest per capita, ranging from a minimum of just over a single hectare to 2,800 ha claimed per person. Communities that had been living in their current location for relatively short periods tended to live further away from their forest claims compared with those that had been living in a single location for hundreds of years. This was due to government policy from the 1970s onwards to resettle communities further downstream and closer to government-provided schools and health facilities (known as *respen penduduk* or 'resettlement populations'). Examples of these kinds of communities include Long Sulit and Paking Lama in Malinau, and Batoq Kelo and Batu Majang in Kutai Barat, all of which were resettled downstream by central government in the early 1980s. These communities also still had small numbers of community members living in the original settlements, nearer to their forest claims.

Figure 5.4 Amount of Land Claim (ha) for all Sampled Communities

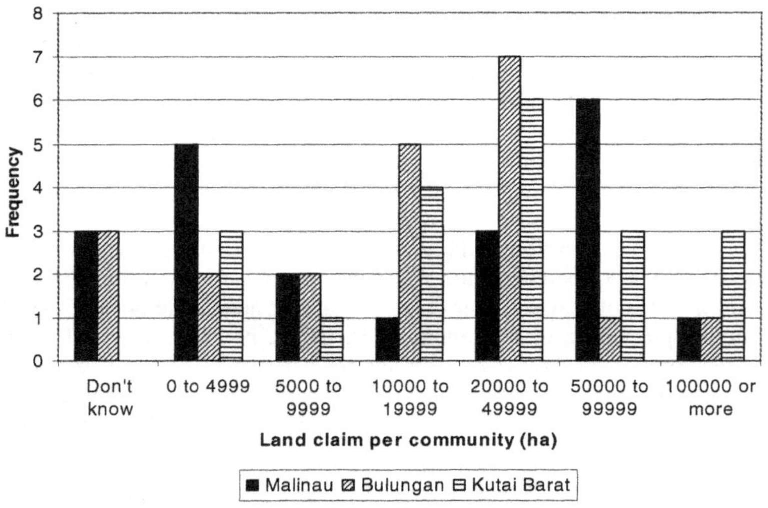

Unlike some communities in Bulungan and Malinau, none of the communities surveyed in Kutai Barat replied 'don't know' to any of the main questions on forest or land claim. Land claim data in figure 5.4 includes all land used for farming and gardens. In many cases, however, data at this level could not be collected and was again, quite patchy for all three districts.

A higher proportion of communities in Malinau (38 percent) have attempted to register land with the local government at some point in the past compared to those in Bulungan and Kutai Barat (24 and 10 percent, respectively), although this tended to be agricultural land or village territory and not forest-land *per se*.

5.6 Community Decision-making: Access to and Use of Community Forest

According to the community responses, decision-making on forest access tended to be restricted to the village leaders in a majority of cases, in all districts. In Kutai Barat, it was usually just one or two individuals, the village head, customary head or both (see figure 5.5).

Given that community-wide processes of decision making over forest access and usage were only present in a minority of the sampled communities, public discussion within the communities about these decisions was a lot more widespread in Malinau and Bulungan than in Kutai Barat. In the latter, public consultation only took place in around 50 percent of surveyed communities, compared with over 90 and 75 percent in Bulungan and Malinau, respectively. Overall, community leaders claimed that the wider community had some general participation in community level decision-making in nearly three quarters of the sample.

Figure 5.5 Community Decision-making over Access to and Use of Forest Resources

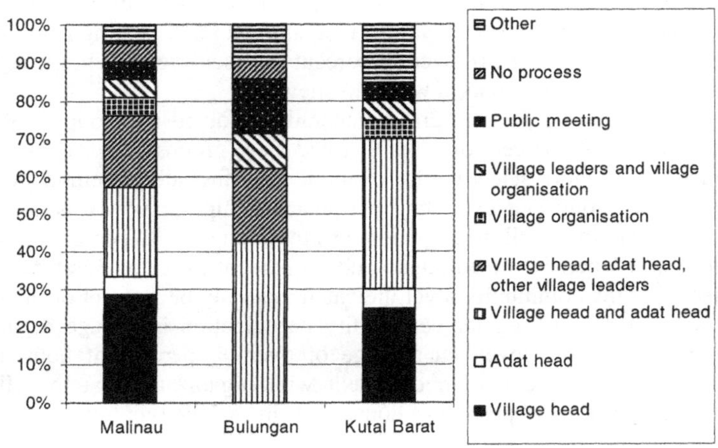

Village organisations were typically linked to village governance in that they normally fulfilled functions that assisted the village leaders in day to day running of community affairs. The commonest organisations were the 'farmers group' (known as *kelompok tani*), the 'representative village committee' (*Badan Perwakilan Desa* or BPD), and the 'organisation of human development' (*Lembaga Pengembangan Manusia* or LPM), all of which focused on the agricultural and social development of communities. These organisations were mandated by the local government and sometimes complemented the more informal, *ad hoc* arrangements of many communities. Some of the farmer groups were originally used as a vehicle for obtaining IPPK permits in Bulungan and to a much lesser extent in Malinau (see chapter 7). Some were used as monitoring teams, while others were actually discussion groups for farmers. Away from the government groups, communities sometimes had groups that managed customary (*adat*) land (*staf adat*), although in most places this was left to the customary head alone. Cooperatives (*koperasi*) were also established to apply for and manage HPHH permits in Kutai Barat, and these tended to have restricted membership.

There was an average of over two organizations per community, with variation from zero to four. Only two communities in the whole sample (three percent) had no organisations whatsoever, and a further 79 percent of the sample have two or more organisations. Over, half of the communities surveyed in Bulungan have three or more organisations, a higher proportion compared to the other two districts.

In numerous cases, it was hard to infer the activities and level of participation in some of these organizations, particularly the government ones. These organisations were in general not viewed as being very important with the exception of some of the farmer groups and the cooperatives. Community respondents were asked how often these organizations met, along a scale ranging from 'more than once a week' to 'never'. In over 90 percent of cases overall, village organisations met either 'less than once a week' or 'never'.

Participation in the farmers group generally involved all groups within a community in over 95 percent of cases that had these groups. Other community organisations were more selective, for example only including members of the dominant ethnic group and excluding peripheral groups or newcomers to the community. In Malinau, full participation was found in only 42 percent of cases, compared with 85 percent in Bulungan and 76 percent in Kutai Barat. Overall, data collected at the community level indicated that only 63 percent of the sample had widespread and regular community participation in village organizations. At the household level, an average of over 30 percent of households claimed to have some regular involvement with organizations at the village level, varying from 0 to 78 percent of households in each community.

5.8 Infrastructure before Decentralization

Prior to making these agreements, a minority of the community sample had relatively little in the way of infrastructure such as roads, health clinics and electricity. 87 percent of communities had schools (at the primary level), 68 per cent had rice mills and 58 percent had electricity at this time. Many communities in Kutai Barat could only be reached by river, even since 1998. While there were relatively fewer external roads in Kutai Barat compared with Malinau and Bulungan, communities in the former had the highest proportions of electricity, bank accounts and rice mills of all the districts (see figure 5.6). For the other indicators, communities in Bulungan tended to have more infrastructure compared with those in the other districts.

Figure 5.6 Proportion of Surveyed Communities containing Selected Infrastructure before Decentralization

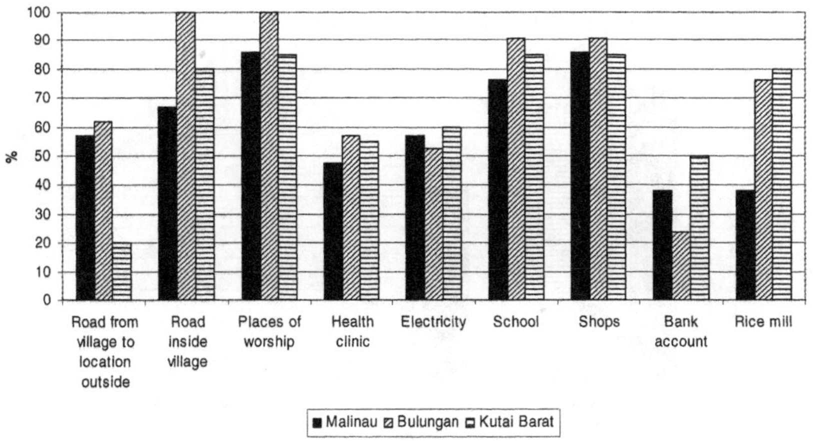

From the household data, an average of 16 percent of households per community claimed to have savings or bank accounts before the onset of negotiations, ranging from 0 to 60 percent across the sample. A further 17 percent and 49 percent claimed to own televisions and some mode of transport (typically a boat), respectively, during the same period. As with bank accounts and savings, there was wide variation in household ownership of televisions and transport from community to community. In Indonesia, the compulsory age of schooling is until the age of 15, of which there was again large variation in the sample. Between 0 and 100 percent of the households surveyed by community had members with post-age 15 schooling, with an average of 49 percent per community.

Almost all households lived in houses characterized by being built out of wood, on stilts and having corrugated iron roofs. Sanitation infrastructure was almost non-existent with the river still being used in many communities as the main place for washing, cleaning and drinking.

5.9 Third Party Assistance

In all three districts, a majority of communities have received some kind of social assistance, either in the form of village development funds or direct developments from the district government (figure 5.7). These developments typically took the form of electricity generation or a new road.

Figure 5.7 Assistance from District Government to Communities in Malinau, Bulungan and Kutai Barat.

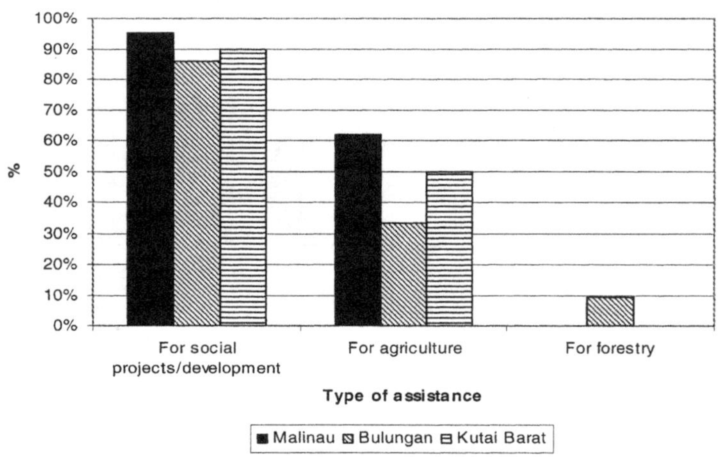

A much lower proportion of communities (39 percent) have received assistance from NGOs (local or otherwise), e.g., in the form of mapping of village territories and development projects, than from the local government. Communities received a greater breadth and depth of assistance in Kutai Barat compared with those in the other two districts. 55 percent of communities received NGO assistance in Kutai Barat compared with 33 percent and 29 percent of those in Malinau and Bulungan, respectively.

6 Community Experiences before Decentralization

In East Kalimantan, the first large-scale commercial logging operations with HPH permits began in 1969-70 and in some places have been in operation for over 30 years since. A total of 60 communities in the sample had experiences of these operations on their territorial claims, before decentralization: 19 in Malinau, 21 in Bulungan and 20 in Kutai Barat. Figure 6.1 compares HPH experiences across the three districts with data based on community level perceptions of when HPH companies entered and began harvesting timber on community territory, and when they finished. In some cases companies were still operating during the survey, and in many cases particularly in Bulungan and Kutai Barat, more than one HPH company operated in community territory, either at the same or at different times.

Figure 6.1 Total Length of HPH Experiences for Sampled Communities

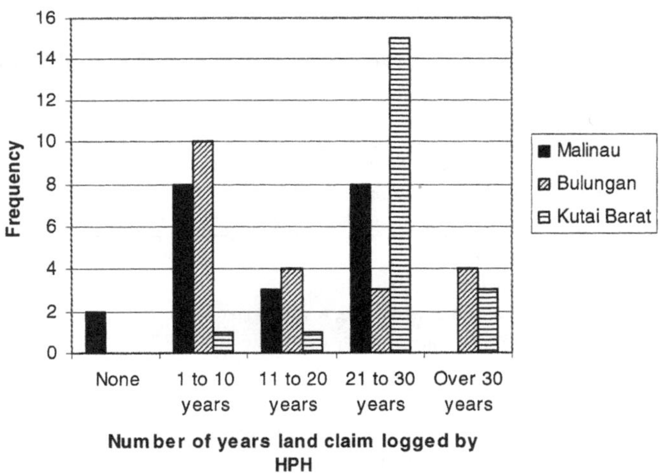

Note These are the total approximate lengths of all HPH operations on land claimed by 60 communities and include those that continued operating post-1997-98.

The most prolonged exposure to HPH activities took place in Kutai Barat where 90 percent of communities experienced these operations for 21 years or more. By contrast, 42 and 33 percent of communities were exposed to operations for this length of time in Malinau and Bulungan, respectively.

6.1 Monetary and Non-monetary Benefits from Logging before Decentralization

Based on responses at both the community and household levels, of the 60 communities, only eight (13 percent) received any kind of compensation or social assistance before 1997-98. No communities in this sample received cash compensation[36]. Instead they typically received building materials such as timber and metal sheeting for houses and village buildings such as churches and meeting rooms, along with agricultural assistance such as farm tools and seeds. These benefits were usually non-negotiable and are listed in table 6.1 along with the proportions of households that perceived these (in brackets). Some villagers were able to obtain work with the companies as chainsaw operators or surveyors, although many HPH companies preferred to bring in their own experienced crews. Precise data on job provision by logging companies to communities before 1997-98 were unavailable.

Table 6.1 Monetary and Non-monetary Benefits of Mechanised Logging before Decentralization (for Communities reporting Positive Benefits)

Community	Benefits before decentralization		
	Monetary (total, IDR)	*Non-monetary (% households perceiving benefits)*	*Length of operations (years)*
1	None	Housing (10)	30 +
2	None	Farm tools (12,5)	10-15
3	None	Agricultural assistance, clean water, ground levelling, housing materials (90)	20-25
4	None	Seeds, development of village lavatories (10)	30
5	None	Mosque (7)	25-30
6	None	Mosque, agricultural assistance, seeds (37)	30
7	None	Mosque, agricultural assistance, seeds (57)	30
8	None	Transportation to fields, agricultural assistance, generator, housing materials (30)	25

Note In all eight cases, village leaders and at least some households stated that benefits were received by logging companies prior to 1997-98. The figures given in brackets denote the percentage of households in the community that confirmed that these benefits were received, either individually by households or by the community as a whole.

Table 6.1 illustrates just how few communities actually benefited from any of the HPH rural development schemes as required by central government, despite

[36] A number of communities in this sample did however receive cash compensation after the 1997-98 changes from their respective HPH companies. These benefits are not considered here.

the relatively long length of logging operations in many villages. For those that did benefit, a lack of precise details on pre-1997-98 benefits makes it almost impossible to undertake a valuation of these. It seems that the benefits were received either so long ago that they were long forgotten or too little for many people to remember at least in comparison to post-decentralization benefits. Responses at both community and household levels were usually vague as to the details of these benefits, and some respondents were confused when asked to differentiate between benefits provided by companies and that provided by the government[37]. Very few respondents had much idea of the scale of HPH operations either. In almost all cases in which village leaders responded that there were some benefits from the HPH companies, only a minority of households responded that this was so with the exception of communities 3 and 7. Overall though, it is clear that benefits were not substantial and in most cases could be termed 'negligible'.

6.2 Protests and Conflicts against Logging Companies before Decentralization

Before decentralization, nine communities in the sample had experiences of negotiating with HPH companies, two of which also participated in activities against the company as well. Neither of these two received any assistance or compensation, while another three did receive something after negotiations. Four of the total of eight communities that received compensation before 1997-98 undertook activities against the company beforehand. Of the remaining 52 communities in this sample, 21 or 40 percent of the sample involved in some level of activity against the HPH companies, ranging from small demonstrations to the confiscation of equipment and timber. Overall, only 18 percent of households claimed to have participated in actions against HPH companies. While actions ranged and details are not available for most of these, it seems that in 84 percent of cases where protests against HPH activities took place, no compensation was offered at least until 1997-98. Thus, protest did not lead directly to

---

37 There were at least two programmes that were supposedly undertaken by central government in poor, rural areas prior to decentralization: IDT (*Impres Desa Tertinggal*, which was an occasional cash infusion from central government to isolated rural communities) and PKMT (*Pembinaan Kesejahteraan Masyarakat Terasing*, a programme that was undertaken by the Ministry of Social Affairs for the 'improvement of the welfare of isolated communities'). Some respondents were unable to distinguish between benefits provided by the two actors, which to some extent confirms the high degree of collaboration between the logging industry and the government during the Suharto era. Fortunately, an analysis of community and household-level surveys enabled distinctions to be made among sources of benefits.

benefits and those companies that did offer compensation sometimes did so without being under any kind of threat from communities.

Those communities that protested were asked for their reasons. There was a plurality of answers, the main one being that the companies were damaging what they perceived to be their customary forest. In addition, some communities commented that they protested because they simply wanted compensation for this intrusion into their territories. Others mentioned that compensation was demanded for specific damage from the logging activities of HPH companies, such as that to community gardens and supplies of drinking water from rivers.

For the remaining communities that did not protest, they were asked for their reasons. An overwhelming majority (86 percent) replied that they were afraid or scared of the company. Many communities it seemed were simply not used to dealing with outsiders at the time, while others specifically mentioned threats by the Indonesian security services, particularly the military and the police[38].

6.3 Environmental Impacts from Logging Companies before Decentralization

Survey respondents were asked about their perceptions of changes (positive, negative, no change) for a number of environmental indicators: river water quality (for washing and drinking); flooding and water levels (can go up and/or down); hunting; collection of forest products, and; forest utilized by community (for agriculture and gardens)[39]. Then, for responses of 'positive' or 'negative', they were asked why they thought there had been a change. Figure 6.2 gives the average proportion of households per community in each district that perceived these negative impacts only, e.g. decline in water quality, increase in flooding etc, that were either partially or completely caused by mechanized (HPH) logging before decentralization. Thus, a 100 percent response would indicate that all households sampled in a particular district had perceived a negative impact say on drinking water quality from HPH logging. The figure also presents the proportion of communities per district giving similar, negative responses (from the focus group meetings). Here, a 100 percent response for a particular district

[38] It is worth noting that many respondents who did experience benefits from companies before decentralization were also afraid of the companies and the power behind them.

[39] In both surveys, respondents were also asked about the impacts on fishing and 'protection forest'. However, these were dropped during the course of the survey. For fishing, respondents and researchers were confused about whether this also applied to man-made fish ponds or just to rivers. 'Protection forest' is a government categorization that many respondents were either not aware of or equated with customary forest.

would indicate that all sampled communities (on the basis of the focus groups) perceived negative impacts from HPH logging.

Overall, the impacts on the quality of drinking water from rivers and flooding were most felt in Malinau and Kutai Barat. The collection of forest products was affected most in Kutai Barat, as was farming and gardens. Approximately 60 percent of community responses both in Malinau and Bulungan indicate negative impacts of HPH logging on farming and gardens. On average, around 10 percent of households in each of Malinau and Bulungan gave a similar response. Many HPH barred communities from using the forest in concession areas and this accounted for the lack of a 'negative' response from households, simply because many households either did not know about or farmed away from HPH activities. In many community discussions on the other hand, the very fact of prohibition was perceived as a negative impact regardless of the environmental consequences of such activities.

Figure 6.2 Community and Household Perceptions of Environmental Impacts from HPH Operations

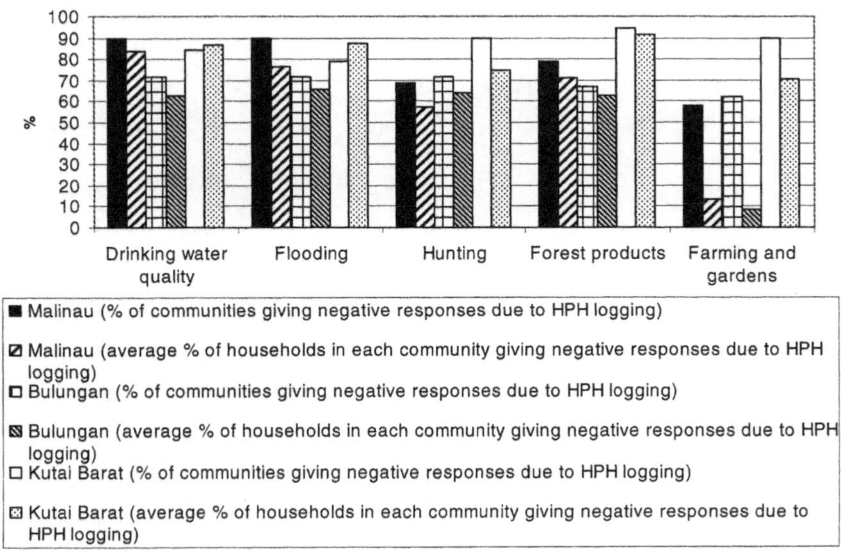

7 Post-decentralization Negotiation Processes

This chapter describes the general sequence of events during negotiations and before a company gained access to a community's forest claim and began logging operations. First, data on previous community knowledge of IPPK/HPHH is presented before the introduction of brokers and their important role in negotiations. Next, generalized negotiation processes are described, with a discussion of the main underlying themes and the main differences observed among communities, particularly with respect to district.

7.1 Prior Knowledge of IPPK/HPHH Systems before Negotiations

From community data, in almost 63 percent of cases, members of the village elite (as opposed to the community at large) were the only ones that had knowledge of the IPPK/HPHH systems and the potential benefits from these, before negotiations commenced. Members of village elites tended to include the village head and the customary head among others. In only 19 and 16 percent of cases, were 'most' or 'some' of the community aware, respectively. Wider community awareness was more common in Kutai Barat than in the other districts.

Over 30 percent of communities learnt about the IPPK/HPHH systems solely from the district government, of which half were in Kutai Barat. Smaller numbers obtained information from the companies (21 percent), other communities (18 percent) and from a combination of the companies and district governments (11 percent). Communities that learnt about the systems from firms tended to have only been made aware at the onset of negotiations.

7.2 The Role of Brokers and Subcontracting

In many cases, the broker was the key liaison between the government, logging contractors and communities[40]. Overall, 68 percent of communities, used brokers from outside the community in negotiations with logging companies, of which over 70 percent were known to the community prior to negotiations, typically through family ties or working relationships. This emphasises the importance of pre-existing networks based on family, work contacts or simply recognition and connections through belonging to similar ethnic groupings. These networks were common to all three districts, although around three quarters of community-level respondents in Malinau denied any pre-existing relationship with their brokers or companies. In the districts of Malinau and Bulungan the

[40] Brokers were called a variety of terms including *putra daerah* ('sons of the area') in Malinau, *pengusaha* ('entrepreneur') and *calo* ('opportunist') in Bulungan, and *raja kayu kecil* ('small timber kings') and *kuasa usaha* (also 'entrepreneur') in Kutai Barat.

use of brokers was more widespread compared with the situation in Kutai Barat, possibly because of differences in timber processing markets and closer contacts between communities and timber buyers in the latter district. As described in chapter 2, timber buyers and contractors in Malinau and Bulungan were mostly Malaysian, due to the close proximity of the enormous timber processing industry in Sabah. In particular, the export ban meant that Malaysians were unable to enter the Indonesian timber market legally, hence necessitating the use of Indonesian brokers. Communities in Kutai Barat tended to deal directly with logging contractors. Where brokers were involved in negotiations, their role was not as crucial as it seemed to be in Malinau and Bulungan.

While not employees of vertically integrated, capitalized logging companies, brokers had the necessary links to contractors, timber buyers, and government officials and of course, local communities. In general, their role was to assist in organizing the community to claim forest areas and to find capital owning contractors. Thus, brokers were located in the centre of a network of actors, all of whom depended on him to develop and arrange all the deals at the production-end of the timber chain: the permits, the buyers, the contractors and the concession areas. This grouping of buyer, contractor and broker is known as a limited partnership or 'CV', as described in chapter 2. Hereafter, they are loosely defined as the 'company'. Thus, the broker acted as the main representative for the 'company' or CV as a whole, although in some cases the contractor accompanied the broker in negotiations with the community.

In table 7.1, generalized steps in negotiations are compared among all three districts. The similarities and differences are discussed in greater detail in the remainder of this chapter.

A small number of brokers dominated deal-making, particularly in Malinau and Bulungan, many of whom were well-known locally, at least to community leaders. Many had previously acted as traders for non-timber forest products (NTFP) or had worked in the logging industry within illegal logging networks. The former saw IPPKs as a lucrative new opportunity, using well-established trading networks in the area, while the latter saw IPPK as a means by which they could legitimise their businesses. Moreover, a number of brokers had links to or had had previous employment with HPH concessionaires in the area as well.

Table 7.1 Generalized Steps Towards Agreement in the Three Surveyed Districts

Steps	Malinau	Bulungan	Kutai Barat
1	Broker scouts out potential IPPK area, makes a 'preliminary'* agreement with contractor and timber buyer	Broker scouts out potential IPPK area, makes a 'preliminary'* agreement with contractor and timber buyer	Brokers/contractors/HPH concession holders (or ex-holders) scout out potential HPHH concession areas
2	Broker approaches district government to come to principle agreement	Broker comes to community or village leader(s) meet(s) broker in Tanjung Selor town or Tarakan town	Broker/contractor comes to community or village leader(s) meet(s) broker/contractor in Melak town or Samarinda
3	Broker comes to community or village leader(s) meet(s) broker in Malinau town or Tarakan town	Broker seeks recommendation from sub-district head	The offer is discussed sometimes only among village leaders and sometimes in a community meeting
4	An offer is made by broker or a demand is made by village leader(s)	Broker seeks recommendation from village head and customary law head, and an offer is made by the broker or demand made by community leaders	Community and company come to an 'preliminary'* agreement
5	The offer is discussed sometimes only among village leaders and sometimes in a community meeting	The offer is discussed sometimes only among village leaders and sometimes in a community meeting	Community members organise into cooperatives, including those selected as potential permit holders
6	If accepted, a letter of agreement is made between the village and the broker. If offer is not accepted, then there are further negotiations	If accepted, a letter of agreement is made between the village and the broker. If offer is not accepted, then there are further negotiations	Community members apply for permits
7	The benefits for the village are formalized in notarized contract	The benefits for the village are formalized in notarized contract, and a farmers group is formed. Copies of identity cards of members are collected for the permit	District head officially approves and issues permit(s) to community members. Negotiations ongoing between community and contractor

Table 7.1 Generalized Steps Towards Agreement in the Three Surveyed Districts (continued)

Steps	Malinau	Bulungan	Kutai Barat
8	Broker takes the letter of agreement from the villagers to district government	Sub-district head, community leaders approach logging contractor and buyer assisted by the broker	Agreement made and the benefits for the village are formalized in a written or notarized contract
9	District Forestry Office provides technical advice related to status of area requested and harvestable timber	District Forestry Office provides technical advice related to status of area requested and harvestable timber	District Forestry Office provides technical advice related to status of area requested and harvestable timber, arranged by broker
10	District head officially approves and issues permit	District head officially approves and issues permit	

Note *a 'preliminary' agreement was typically oral and not as binding as the final agreement.

7.3 First Steps: Scouting for Forest Areas and Initial Contacts

First, potential, commercially viable IPPK/HPHH concession areas were scouted and the communities claiming these areas were identified. Brokers usually took the initiative to find potential concession areas, although local contractors have been known to do this as well. In Bulungan, areas were being scouted by mid 1999; in Malinau by late 1999 to early 2000, and in Kutai Barat by early 2000 (step 1 for all districts in table 7.1). Many of these areas were located in or near established HPH concession areas. Secretly, some HPH operators worked with brokers and local contractors to secure prospective IPPK/HPHH concessions either in or near established HPH concessions areas, in order to expand their concession areas. IPPK/HPHH concessions were placed inside HPH areas in 60 percent of all cases, particularly in Kutai Barat. For example, the HPH company PT. Sumalindo had a secret profit sharing arrangement with the contractor, PT. Fahmi, which made an agreement with the community of Batu Majang.

Even at this stage, it seems likely that the broker would have some kind of preliminary arrangement with a timber buyer and a contractor. This was important, particularly in Malinau and Bulungan, because timber buyers and contractors were responsible for providing cash 'up front', for example, for the payment of 'donations' for arranging permits in government agencies. The brokers alone only very rarely had this kind of money up front.

In scouting out potential concession areas, the brokers were not only relied upon to have good knowledge of the commercial viability of local forest, but also knowledge of local power relationships and community land claims. Community land claims had never been formalised and hence, in setting up deals, the brokers had to know who to approach on the community side. In addition, brokers were relied upon to have good local knowledge when checking the relative credibility of territorial borders before making a formal approach to communities. On the other hand, when communities themselves searched for potential deals they were relying on others to recognise their respective land claims. The credibility of forest claims appears to have been dependent on a number of factors, including history and previous usage. For example, a number of communities in the sample had forest claims in areas where they no longer reside, while others made new claims in areas also claimed by other communities. For example, Long Simau, Long Sulit and Batu Majang were resettlement communities that still maintained populations in their original settlements. These old settlements were hundreds of kilometres from where the majority of the members moved in the 1980s. Ultimately though, it was always the district head who had the final say on the credibility of these claims.

Different to Bulungan and Kutai Barat, brokers in Malinau usually approached the district government first and made some arrangements before approaching communities (step 2). Thus, the broker decided in advance with district government officials about certain elements of the community-company agreement such as the size of the IPPK concession area, any logging borders and logging rules. Nevertheless, some communities still discussed (and even negotiated for) some of these elements, particularly since there appeared to be a lot of ambiguity over the precise nature of these in the IPPK regulations (see chapter 8).

Preliminary contact was made with the community controlling access to the forest area, although brokers and contractors never entered negotiations with those that were already in the process of making a deal with somebody else (step 3 in Malinau; step 2 in Bulungan and Kutai Barat). Thus, there was first-mover advantage in securing as many deals as quickly as possible, which would explain the rush to secure agreements and arrange permits from 2000 until 2001. That these brokers dominated their respective areas of influence meant there was little competition with other brokers for making agreements. Approximately 35 percent of the sample had approaches from other brokers or contractors during the course of negotiations, although alternative negotiations did not take place at the same time as these. The rush to secure permits and agreements meant that brokers were holding numerous options on potential concession areas, but did not always have the capacity to take them up. But these options still held value.

Where brokers were attempting to arrange deals alone, they would also bring the contractors and communities together in what was sometimes known as 'socialisation' (*sosialisasi*). This involved at least one face to face meeting between the community and the contractor to build up trust and demonstrate to the community that the broker was serious about making an agreement. In a small number of cases (19.4 percent), gifts were brought to these initial meetings by the brokers and contractors. Most of these cases were in Malinau. Usually, it was the broker who continued to liase with the community, not the contractor.

7.4 First Meetings, Negotiations and Participation

Soon after the initial contact, the brokers and/or contractors initiated negotiations with communities for access to forest areas, or the communities initiated negotiations with the brokers/contractors. In the latter case, the broker/contractor then surveyed the commercial potential of the proposed area. In 63 percent of cases, the initial discussions were held in the community, followed by towns and cities (24 percent) and other villages (13 percent). Almost all the initial discussions that took place in a different community were located in Malinau, which mainly reflected the shared nature of some of the agreements there. In Kutai Barat on the other hand, 35 percent of initial discussions took place either in Samarinda, or in the other major city in East Kalimantan, Balikpapan.

During these initial discussions, typically, offers and/or demands were exchanged by the two parties, which then led to further discussions, meetings and negotiations. In these first meetings, the offers or demands were made in terms of financial benefits (advance payments, fee per m^3), jobs and social developments. Companies made promises in 30 cases (48 percent), with most of these in Bulungan and Kutai Barat, during the first meetings. Moreover, during the same meeting, an agreement was also made in a further 13 cases (21 percent), which were predominantly to be found in Kutai Barat. Before a 'final' agreement was made, 'further discussions' were held between the company and the community representatives in 41 communities (66 percent). Some issues such as for example the nature of jobs for community members would be discussed and resolved some time after the main points of the agreement had already been agreed upon. That said, only 36 communities (58 percent) thought that they had 'enough time' for considering the agreements, a figure that falls to less than 40 percent of communities in Kutai Barat.

From the community side, there were usually individuals or even a single individual who met the broker first and negotiated on behalf of the community. These individuals tended to control access to the forest. Some of the most powerful individuals involved represented families known as 'inheritors' (or *pewaris*[41]), or were customary law heads for an entire watershed (*kepala adat be-*

[41] The term pewaris normally referred to a powerful local family, although community members sometimes used it to refer to a single person such as the head of one of these families. Their land claims were typically based on inherited rights to birds nests, usually to be found in caves, and sometimes the timber around them. These private 'rights' were inherited either

sar). These individuals sometimes had strong land claims of their own. There were far more of these kinds of local, powerful people involved in initiating or helping to make deals on behalf of communities (in 85 percent of cases) in Kutai Barat than in Malinau or Bulungan. Sometimes, these individuals also undertook the role of broker for the few communities in Kutai Barat that were observed to be using them (see chapter 8). Typically though in most cases and in the absence of a single powerful family such as the *pewaris*, the village head and the customary head, and sometimes other prominent community members or elders would represent the community in negotiations. Thus, the majority of negotiations surveyed involved a few well-placed people dealing with the broker, at least during the initial meetings. Most community members were only involved in negotiations indirectly, e.g. through consultations, and their impact on decision-making was not always easy to discern.

Using community data, figure 7.1 illustrates who from the community met with the companies first and in the following negotiations. Where there are only select individuals involved such as village leaders or the *pewaris*, this is to be differentiated from cases where 'some' or 'all' of the community were involved. In the latter cases it can be assumed that there was a greater element of openness and non-exclusivity relative to cases where only well-placed members of the community involved in the processes of discussion and negotiations. For each district in figure 7.1, the second set of columns describes who else, if anyone was brought into the negotiation process once it had already been initiated (steps 4 to 5 in Malinau; steps 5 to 6 in Bulungan; steps 3 to 6 in Kutai Barat). Thus, these data should be interpreted as additional to those illustrated in the first set of columns. Where 'some' or 'all' community members participated (whether in the 'first meeting' or 'all negotiations'), this was inclusive of individuals such as the village head etc. In all three districts, a small group of well-placed community representatives tended to be the only ones involved in the initial meetings, but following this other members were typically brought more into the negotiation process. This process was more marked in Bulungan and Kutai Barat than in Malinau. Note that figure 7.1 only reflects meetings in which company representatives were directly involved in meetings and discussions, and not internal community meetings about the proposed IPPK/HPHH agreement.

---

from the former Dutch colonial authorities or the former sultanate and were widely recognized by many people in the local area, including local government officials. Land claimed by pewaris was called tanah warisan ('inherited land'), which usually formed part of a community's land or hutan adat (customary forest) claim. In many cases, the pewaris were more powerful than the village head in a single community because they tended to represent ethnic groupings and have family spread out among several villages. Thus their land/resource claims transcended village boundaries. There has been an evolution from a cave heir (*pewaris gua*) to a heir in a broader sense of land claim (*pewaris tanah*) because this is now seen as a powerful means of achieving recognition for all claims.

The negotiations in which 'all' community members were directly involved tended to be in communities that were relatively small comprising perhaps a few dozen households, e.g. Paking Lama and Punan Setarap. In larger communities the gap between the key decision makers and the rest of the community tended to be greater than in smaller ones. For example, in Malinau Seberang, a community with approximately 1,500 members, many households complained that they were never fully involved in or consulted about the proposed IPPK. Joint arrangements involving more than a single community accounted for over a third of agreements surveyed in Malinau and similar to larger communities, sometimes led to less household participation in negotiations. For example, most of the semi-nomadic Punan groups surveyed were involved in joint arrangements with non-Punan, 'sponsor' communities. These 'sponsor' communities tended to take the lead in IPPK negotiations and sometimes did not involve the Punan in the initial meetings with the broker, e.g. the community of Punan Bengalun did not negotiate directly with the broker but indirectly through the village leaders of Sesuà. Thus, in many cases, the wider community had no direct involvement in the negotiations. Typically, they only met the broker or contractor for the first time, once the deal had already been made, during a 'socialisation' visit.

Figure 7.1 Community Participation in First Meetings and Further Discussions with Company Representatives.

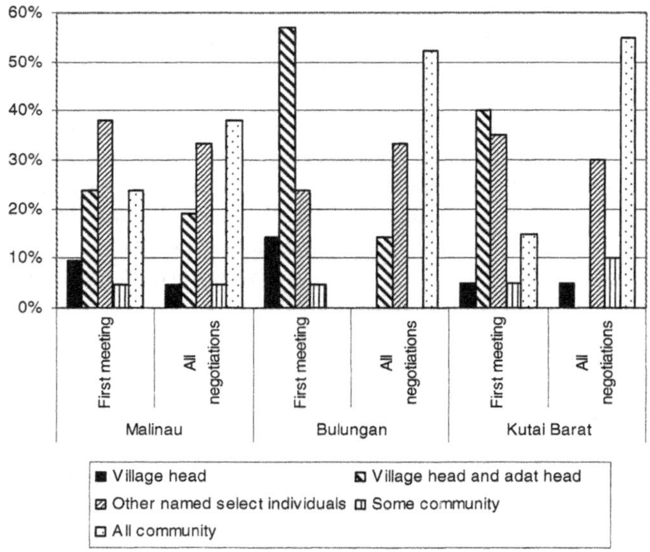

Communities usually held public meetings to discuss the proposed agreement, in which there were no company representatives present. Community meetings and consultation apparently occurred in 57 communities (92 percent). From household data, an average of 86 percent of sampled households per community stated that there were such meetings, with a further 69 percent actually participating in these. While there was relatively little variation from district to district, communities in Malinau had the highest rates of household participation in public meetings. Non-participation was typically due to people being busy, either in or outside the community, or due to 'not being invited'. The last answer would obviously question just how public and inclusive some of these meetings actually were.

During negotiations there were sometimes members of other communities involved, particularly when agreements were to be shared between or among other communities, as was the case in Malinau. Moreover, there were some cases where local government officials were involved. Communities in Bulungan had more local government involvement than any other. NGO involvement was minimal in all districts.

7.5 Formal Procedures for Application of IPPK/HPHH Permits

There were some formal procedures that were unique to each district, with some in Malinau taking place in advance of the initial contact between the broker and the community as described earlier. Prior to actual negotiations (but after the initial contact) in Bulungan, the broker typically sought recommendations from the sub-district government (step 3 in table 7.1), particularly the sub-district head. The sub-district head helped to make the proposal and on the requirements and conditions that had to be complied with in order to obtain the permit, at the sub-district government office. This included gathering the identity cards of individuals in the community and organizing a farmers group. The broker was formally required to obtain recommendations and signatures from community members for the transfer of customary forest from the village and customary heads at the same time as negotiating an agreement between the parties, before proceeding to the district forestry office.

In Kutai Barat, the procedure was slightly different again. At around the same time that negotiations and discussions were ongoing, community members organised themselves or were organised by brokers into cooperatives (*koperasi*, step 5 in table 7.1). This was usually undertaken after a preliminary agreement was made between the parties (step 4). Similar to the farmers groups in Bulungan and to a much lesser extent, Malinau, these cooperatives were set up for the purpose of applying for permits. However, unlike the other two districts, a significant proportion of communities in Kutai Barat actually applied for and obtained permits in their names. In all three districts the size and ownership of the permits are decided in advance of any formal agreement between the communities and companies. However, the crucial difference is that permits were only issued by the district government in advance of the formal agreements in Kutai Barat. The formal agreement was required by the district governments in Bulungan and Malinau before any permits were issued.

7.6 Decision-making and Agreements

Community level responses suggest that 32 communities (52 percent of sample) had a 'public vote' on whether or not to accept the final agreement. A further four communities (6.5 percent) claimed some form of 'public consultation'; while the remainder responded that decisions were taken by members of the village elite. By contrast, figure 7.2 shows the proportions of households in each community that stated that there was a public vote of some kind on whether or not to accept the agreement. Communities in Malinau had the highest average proportion of households stating that decisions were made by voting (63 percent). Only a fifth of households on average per community in Kutai Barat stated that decision-making was truly democratic, a figure that was almost matched by the response rate in Bulungan (29 percent).

Figure 7.2 Household Data on Decision-making with Respect to IPPK/HPHH Concessions

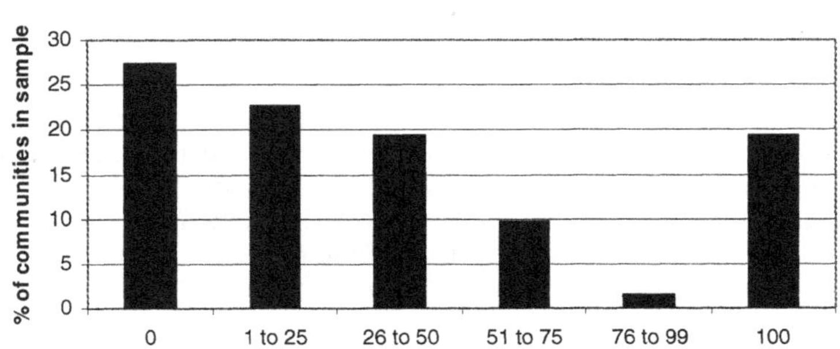

% of households per community answering that there was a public vote

Figure 7.2 shows that in 12 cases (19 percent of the sample), decisions were almost certainly made 'democratically', with all households stating that this was the case. In one community, a high majority of households (between 76 and 99 percent) stated that there was a public vote, followed by six more with relatively small majorities (51-75 percent) of households stating that there had been a vote.

Thus, in total there were only 19 communities in which majorities of households said that they had a say in decision-making with respect to HPHH/IPPK concessions, in comparison to the 32 community responses. Instead, a majority of households in 69 percent of communities in the sample stated that decisions were typically made by village leaders, perhaps with some consultation during public meetings.

In 92 percent of cases, community leaders stated that the community as a whole was made aware of the contents of the final agreement through public meetings even in cases where all decisions were made by village leaders. However, it seems that participation in public meetings did not result in full awareness of the decisions being made or the details of what was being decided. An average of 50 percent of households per community was aware of the decision to accept the agreement when it was originally proposed, although only 38 percent actually knew the contents of the agreements. Awareness of decision-making was highest in Bulungan (60 percent), followed by Kutai Barat (55 percent), and lowest in Malinau (37 percent). Conversely, knowledge of the agreements was highest in Malinau (46 percent), followed by Kutai Barat (41 percent) and lowest in Bulungan (30 percent). This suggests that while households in Kutai Barat and Bulungan knew about the decisions that were made with respect to IPPK/HPHH agreements (even if they themselves were not directly involved), they did not know what these decisions actually resulted in, i.e. the contents of the agreement. In Malinau on the other hand, households had slightly better knowledge of the agreements made but not how they were made.

Soon after in almost all cases, a formal letter of agreement was written up, sometimes leading to a notarised contract (known as *akte notaris*) between the two parties. While all communities surveyed in Malinau made notarised agreements with the broker, these were much less common in Kutai Barat (see chapter 8).

7.7 Permits

In Bulungan, the agreement between the community and the broker was less important for the issuance of a permit. Instead, the rules stated that for a permit to be issued, there needed to be a written explanation how 'activities' were to be conducted ('intentions') by the permit owners and the farmers group. The implication was that the IPPK permit as a 'forest clearance' permit was to be used as a means of plantation development, although it quickly became a *de facto* timber harvesting permit. Furthermore, this explanation was be taken along with the written recommendations from village leaders and sub-district head, copies of identity cards, map of location of proposed 'work area', technical r-ecommendation from the Forest Office and other papers to the office of the district head. In Bulungan it was clear that the copies of identity cards belonging to community members were necessary for allowing a transfer of customary forest to the broker. The community members in question were only told by brokers that their identities were necessary for starting the farmers group, not for the transferral of forest use rights despite the fact that the whole purpose of the IPPK was to provide these rights to communities in the first place. In other words, there appeared to be an element of deception in this transferral of use rights in Bulungan.

In Malinau, after making an agreement, the names of the families along with their signatures would be taken away from the village by the broker to the office of the district head to obtain the permit. When the system was originally established in 2000, the rules stated that each family that applied for a permit was entitled to utilise 100 ha. Similar to Bulungan, the community members themselves did not apply for the permits directly, but their names were used so that the broker could claim ownership of the permit (see figure 7.3). One of the reasons why this might have occurred was because it was probably too expensive for many villagers to afford the cost of permit application, although similar, relative material poverty did not stop community members from applying and owning permits in Kutai Barat. It is more likely that the brokers wanted to own the permits partly because they were the ones directly dealing with the timber buyers and contractors and partly because they wanted to retain the option of selling it on at a later date if it was not immediately operationalized.

In Kutai Barat, the system worked the other way round. Permits were usually obtained before the company made a formal agreement with the community. Also, in 17 out of 21 communities, community members and village leaders approached the district government, applied for and owned permits. Similar to Malinau, one permit for 100 ha was allowed per family. Cooperatives were formed in many communities to pool 100 ha permits and create larger concession areas that were more commercially attractive to prospective contractors. Pooling may also have allowed for a sharing of permit application costs, although in some cases it was obvious that only those community members with access to savings or credit could feasibly apply for permits unless they were able to negotiate the borrowing of these costs, set off against future timber revenue.

A single permit was in theory only for 100 ha, although respondents in Malinau and Bulungan would always talk about 'the permit' regardless of the actual size of the concession area. In Kutai Barat on the other hand, this distinction was maintained in all sampled communities. So, where a community had for example negotiated a concession of size 1,500 hectares, this would always be described as 15 permits (or more commonly 100 hectare blocks known as *petak*). In Malinau and Bulungan, none of the community respondents were able to give the cost of permit application for their respective districts, whereas half of the respondents in Kutai Barat gave a cost for application for permits.

Figure 7.3 Location of Permit Owners

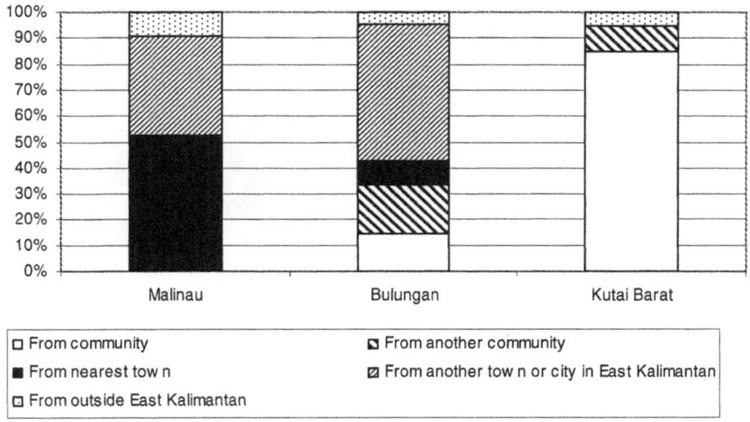

Many of the actual agreements between communities and companies gave areas a lot larger than that listed on the permit, the permits for these agreements would simply be extended every year to accommodate further operations (see chapter 8), at least until district governments stepped in to cancel all IPPK/HPHH permits during 2002-03 (see chapter 9).

8 Contractual Outcomes from IPPK/HPHH Negotiations

8.1 The Companies, Agreements and Market Structures

In Malinau, 15 first agreements were surveyed among the 21 communities sampled there. Five of these agreements were shared between two communities each, one among three communities, mainly due to overlapping forest claims and historical cooperation among these communities. The remaining nine surveyed agreements were independently negotiated by nine communities. Malinau's agreements were individually negotiated with nine different brokers and two different contractors. In Bulungan, 21 communities were sampled, and all had individual agreements that were individually negotiated with 12 different brokers and three contractors. In Kutai Barat, 20 communities were sampled. Two of these originated from the administrative village of Ujoh Bilang and three from the administrative village of Mamahaq Besar. Together, there were 19 agreements, one of which was shared between two communities. These agreements were individually negotiated with 12 different contractors and three brokers.

Figure 8.1 'Companies', Agreements and Market Structure

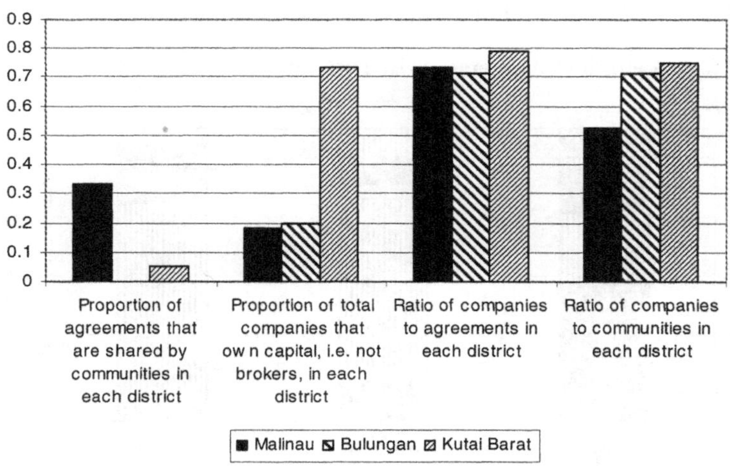

Note Total: 62 communities, 55 agreements, 41 companies. 'Companies' used in the ratios are the sum of contractors and brokers in each district.

Figure 8.1 illustrates the ratios of 'companies' (sum of contractors and brokers) to the number of agreements in each district, and 'companies' to the number of communities in each district, as indicators of district market structure. On both scores, Kutai Barat rates slightly higher than the other two districts, suggesting that there may have been relatively more competition among companies although this does not rule out price-fixing cartel behaviour by the companies in the area.

There was certainly less variation in negotiated fees where a single broker was involved in making a number of deals in the same area (see section 8.4). For example, Batu Lidung, Sesuà and Punan Bengalun in Malinau along with the communities of Sebiday, Kujau and Limbu Sedulun in Bulungan all negotiated agreements with the same broker, all with the same fee of Rp. 20,000 (USD 2.22) per m^3.

8.2 Length and Types of Agreement

A majority of first agreements (93 percent) in all three districts were made in 2000 and 2001, with the remainder made in 1999 or 2002. From figure 8.2, it can be seen that around half of all agreements in Malinau had open-ended time frames, a situation that can be seen for a fifth of agreements in Bulungan and not at all in Kutai Barat. Many of these 'open' agreements were for larger concession areas that required constant permit extensions. A majority of agreements in Bulungan and Kutai Barat were for 12 months only.

Figure 8.2 Length of Agreements by District

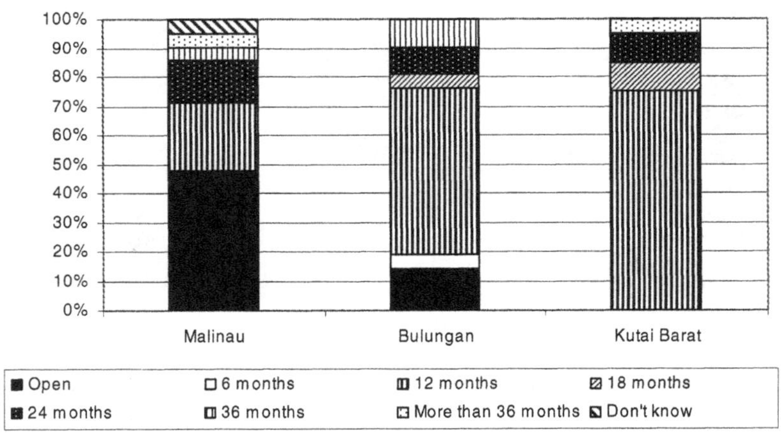

Figure 8.3 describes the breakdown of agreements by type. In all areas, the agreements tended to be a mixture of written and oral terms. For example, in Malinau all agreements were notarised although community employment tended

to be negotiated orally outside the notarised agreement. While Kutai Barat had the lowest proportion of notarised agreements, written contracts were still the norm as in other areas.

Figure 8.3 Proportions of Different Types of Agreements by District

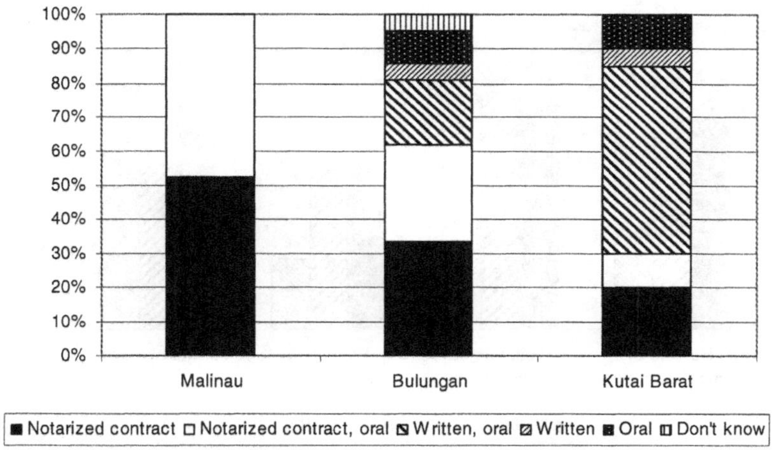

8.3 Terms in the Agreement

Contractual concession areas varied from 200 to 20,000 ha, although these were on average higher in Malinau (6,600 ha) compared to the other two districts (3,268 ha and 2,511 ha in Bulungan and Kutai Barat, respectively). This is perhaps unsurprising given the higher number of open-ended agreements that were negotiated in Malinau.

Overall, over 80 percent of agreements contained terms relating to the borders and locations of IPPK/HPHH concession areas, of which there appears to be relatively little variation from district to district (see figure 8.4).

The IPPK regulations from the district governments in Malinau and Bulungan refer explicitly if not in great detail, to the mandatory development of plantations as a responsibility of the permit holders once the 'working area' had been cleared. This explains why plantation development features heavily in agreements from these districts, and also why some agreements did not contain this requirement: some communities already knew about the government regulation and hence did not think that it would be necessary to include a mandatory government rule in their own agreements. Since HPHH regulations contained no such requirement, there were few promises of plantation development in Kutai Barat.

Figure 8.4 Proportions of Agreements Containing Terms on Borders and Locations of Concessions, Plantation Development and Employment

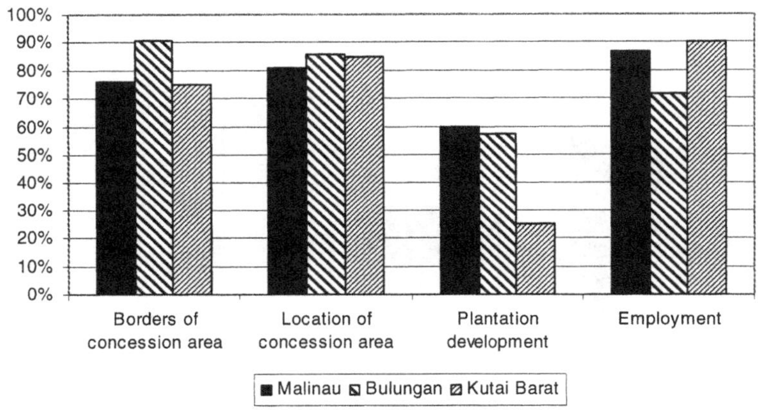

From figure 8.5 on environmental provisions, 'reforestation' requirements were only present in 20-25 percent of agreements in Malinau and Bulungan, and close to 65 percent of agreements in Kutai Barat. Although HPHH regulations followed HPH regulations in that companies were supposed to pay a reforestation tax in lieu of actually replanting anything, this however, did not stop communities in Kutai Barat from including reforestation terms in their agreements. There was a 100 percent overlap between those communities asking for 'reforestation' and 'plantation development' in this district. Typically, a community might ask the company to reforest the logged over concession area and to establish a small plantation of say coffee or rattan nearer to the village. In Bulungan, only half of the communities asking for reforestation also asked for plantations, while there was an overlap of around 80 percent in Malinau. A third of all agreements in Bulungan and Malinau contained no requirements relating to logging behaviour by firms, compared with a fifth in Kutai Barat.

As shown in figure 8.5, diameter limits were popular. These were imposed to prevent firms from removing young trees and hence allow for forest regrowth and prevent further forest degradation. They were typically set at around 50 cm. The restriction to certain species was to prevent firms from removing trees that were utilized (economically or not) by community members. The location of concession areas was typically framed in terms of distance from the nearest river, in order to prevent soil run-off and the silting of rivers. 'Timber production targets' were not very common, and are included here as a 'logging rule' because production schedules were sometimes used to monitor the harvesting rates of companies and ensure that certain areas were not over-harvested, such as those near rivers. Similar to reforestation and plantation developments, other, limited environmental guidelines were issued by district forestry offices. Sometimes those communities that were aware of these had them incorporated in their agreements with companies as well while others insisted on their own guidelines, regardless of what was issued by their respective district governments.

Figure 8.5 Proportion of Agreements Containing Terms Relating to Various Logging Rules

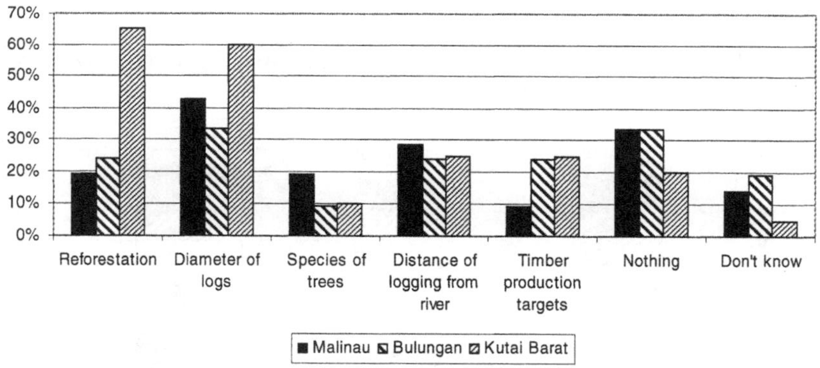

In figure 8.6, it can be seen that over half of the agreements in Kutai Barat contained little or no advance payments compared with those in Malinau where every agreement paid out something in advance of logging operations. These advance payments (known as *uang potong* or *uang pinjam*) were 'borrowed' in advance of production and hence were taken out of future fee payments. In only one case in Malinau was an advance payment not set off against payments from future timber production. However, in Kutai Barat, even though money was rarely paid out in advance, permit costs were sometimes 'borrowed' against future fee payments (known as *dana pinjaman*), and recorded as a production-based sum in agreements (see chapter 10).

Overall, 39 communities (63 percent of the sample) negotiated contracts containing some social developments. Communities in Malinau negotiated contracts that contained more social developments than those from the other two districts (figure 8.7). The commonest items tended to be village buildings such as churches, meeting rooms, village offices, health clinics and houses, in addition to the building of roads and ground levelling, either for houses or sports pitches. Sometimes these promises came in the form of the provision of building materials or simply cash for the purchase of building materials. Most of the time, however, communities fully expected the companies not only to provide the building materials but also to construct the social developments as well. Around 40 percent of Malinau's agreements contained obligations for the provision of utilities such as clean water and electricity, and a third contained obligations for a wide range of other benefits including transportation, agricultural assistance, televisions and satellite dishes. In shared agreements, non-monetary benefits were usually shared. From figure 8.4, 52 communities (83 percent) negotiated for employment with the company. As noted in chapter 7, these promises were mainly oral and in numerous cases were only vaguely defined.

Figure 8.6 Proportions of Agreements Containing Advance Payments of Varying Sizes

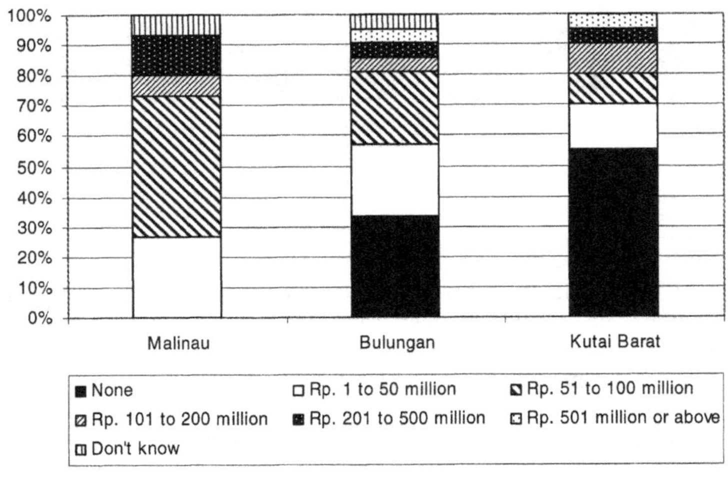

Figure 8.7 Proportions of Agreements Containing Obligations to Social Developments.

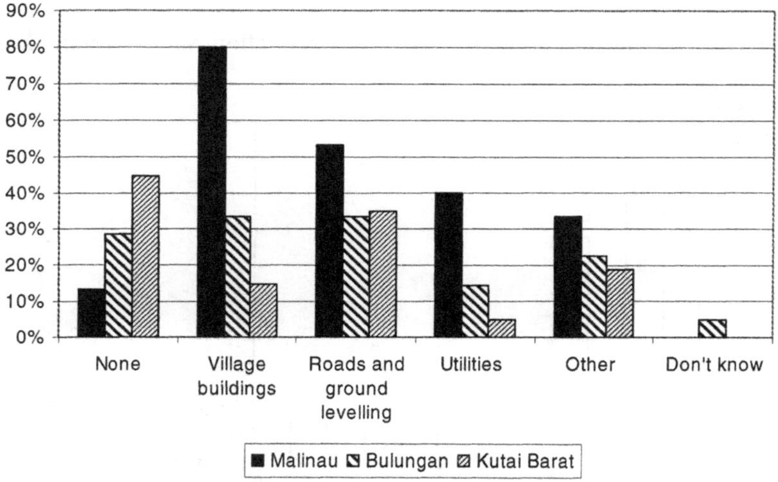

8.4 Production-based Fees

In addition to almost all advance payments, many of the non-monetary developments listed in figure 8.7 were taken from fee payments made to the communities, although apparently not all communities were informed that this was how the benefits schedule was supposed to work. These fees were the key benefit contained within all agreements and all were quoted as a payment per m^3 of timber production. Figure 8.8 illustrates the promised fees. The figures presented here are net of any permit costs made, which worked out at a maximum of Rp. 20,000 per m^3 (USD 2.22) where these had not been explicitly covered in the fee schedule. This is to allow for direct comparison with other communities in the sample in which brokers had already taken out permit costs before making agreements with communities. The average fees in Malinau, Bulungan and Kutai Barat were Rp. 27,125 (USD 3.01), Rp. 32,452 (USD 3.61) and Rp. 73,900 per m^3 (USD 8.21), respectively. Kutai Barat had the greatest range, from Rp. 30,000 to 130,000 per m^3 (USD 3.33 to 14.44). However, it is worth noting that only a few agreements in Kutai Barat contained social developments and the value of provision tended to be included in the overall fee structure. Also, in many cases reforestation was to be paid for out of these fees. There were also some payments that were explicitly contained within the agreements in Kutai Barat that while in existence in the other districts were not included in the fee schedules there. These are considered further in chapter 10.

Figure 8.8 Promised Fees, Rp. per m³ of Production Net of Permit Costs

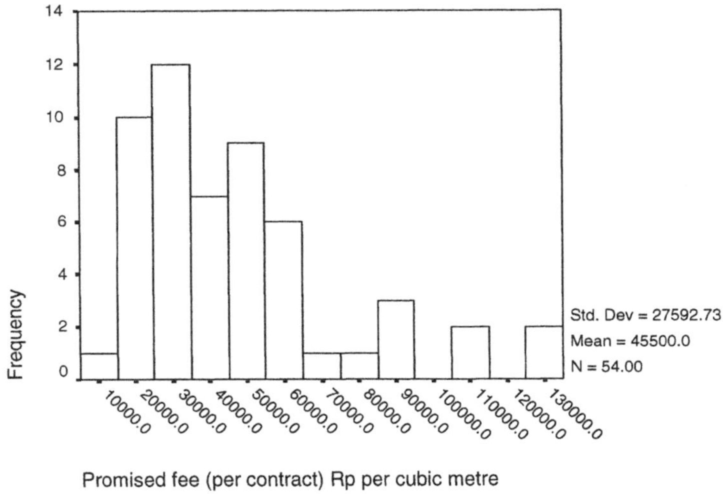

Note Fees for all agreements included except for Long Sulit due to missing data

8.5 Community Expectations and Renegotiations

In some cases, particularly the first communities to involve in IPPK/HPHH negotiations, community expectations were either non-existent or very low. Long Adiu and Punan Adiu for example, only asked for a fee of Rp. 5,000 per m^3 (USD 0.56) due to having had previous experience of receiving compensation of Rp. 3,000 per m^3 (USD 0.33) from a HPH that was operating on community territory. In fact, the broker negotiating with these two communities offered Rp. 20,000 per m^3 (USD 2.22). Given that in theory he could have offered anything between Rp. 5,000 to 20,000 per m^3, this could be more evidence of local brokers profiling themselves as being concerned with community welfare. Hence, this could be interpreted as an investment in political capital in the formation of networks with government officials. In most cases however, communities asked for a lot more than was offered and eventually accepted by community negotiators. For example, Malinau Seberang originally asked for Rp. 70 to 80,000 per m^3 (USD 7.78 to 8.89) and eventually accepted Rp. 40,000 per m^3 (USD 4.44).

The existence of written agreements and even ones that had been notarised should not abstract from the fact that agreements were fluid; many changed or became subject to renegotiation. At least 27 out of the 55 agreements surveyed (49 percent) changed, the majority involving some kind of fee change perhaps with a trade-off with other kinds of benefits, such as social developments. A number of these changes occurred as a consequence of firm non-compliance (see chapter 9), although many changed simply because of changing community preferences. For example, Long Simau renegotiated its agreement so that all social facilities were traded for a fee increase, to Rp. 50,000 per m^3 (USD 5.56). This occurred after the agreement had already been made but before the community received the first payment. This was due to the concern that the facilities built by the company would only be experienced by those living in one of two locations constituting Long Simau. Some communities that had originally negotiated higher fees than the sample average were eventually renegotiated down. These communities had renegotiations imposed by firms that complained of low timber prices, thus impacting profitability, e.g. Keliwai had to renegotiate from its original fee of Rp. 70,000 (USD 7.18) down to Rp. 40,000 per m^3 (USD 4.44).

9 Actual Outcomes from IPPK/HPHH Negotiations

9.1 Length of IPPK/HPHH Operations

By the time of the survey between September 2003 and January 2004, all of the first agreements that had not already finished were forced to cease operations, including those with open-ended durations. The main reason, as described in chapter 2, was that after constant pressure from central government, district governments stopped issuing permits including extensions to existing ones. The district head of Malinau cancelled all IPPK permits in April 2003 and the district head of Bulungan issued a similar decree in December 2003. In Kutai Barat, all HPHH were cancelled as of end 2002 by the district head. However, it was observed in some places in Bulungan and Kutai Barat that some operations may have simply been scaled down and continued without permits. Figure 9.1 gives the length of all IPPK and HPHH operations in this survey.

Figure 9.1 Length of Company Operations

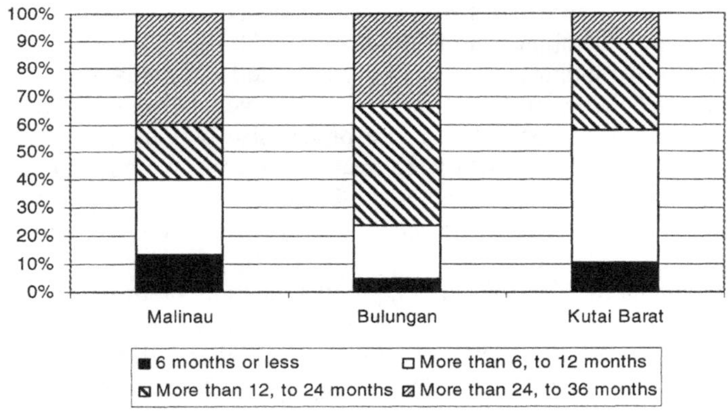

9.2 Area Logged and Log Production

Data for the area logged by the company was received from 56 (90 percent of sample) of communities. Moreover, to qualify this data community respondents were asked whether the area logged was less than, the same as or more than the area agreed. Respondents replied 'less' in 32 cases, which is perhaps not surprising given the many long-term agreements that were made, particularly in Malinau. Log production data was available in 40 communities (65 percent of the sample), almost half of which were located in Kutai Barat.

Data on log production and area logged was mainly derived from communities' systems of checking or enforcement, which tended to revolve around counting logs in loading areas and log ponds. While almost all communities checked the companies' activities at some point, there was a huge variation in systems of checking and monitoring firm activities, thus making a direct comparison very difficult. In most cases, community members who undertook such checks actually worked for the company as well, although some seemed to receive money from the community rather than from the company directly. Only a small number of communities claimed to undertake a wide range of checking activities, including company activities, borders, replanting etc, in addition to log production.

9.3 Monetary and Non-monetary Benefits Received

Data for monetary payments received and non-monetary benefits provided by companies in the duration of the first agreement was computed for all sampled communities, using a combination of community and household responses. The community figure for monies received typically included the advance payment as well, and was available for 48 communities (77 percent of the sample). These figures were checked against the structure of fee distributions and household data for cash received. Where data were unavailable from the community surveys, household data were used alone in combination with the fee and distribution schedules. The fee schedule also accounted for monies not given directly to households.

Combining data for area logged, log production and monies received gives fee payments per m^3 of log production, which allows for a direct comparison of benefits across communities as well as with promised fees. Where unavailable, approximations of log production and/or area logged were made from the length of company operations. The levels of fees received per m^3 of log production, net of permit costs, can be seen in figure 9.2.i. Furthermore, communities and households were asked what other benefits (social developments, etc) were provided by the companies. On the basis of data collected from the value of social developments contained within some contracts and households, these non-monetary benefits were valued and added to monetary benefits (see figure 9.2.ii). Thus, non-monetary benefits were valued narrowly and crudely as the cost of provision, and without further consideration of any other values, e.g. the future flow of benefits from the education of community members who use a school constructed by a company.

The mean level of monetary provision alone is Rp. 31,834 (USD 3.54) per m^3 of log production, and for financial provision plus social developments, Rp. 32,789 (USD 3.64) per m^3. The minimum and maximum levels for both measures are identical: Rp. 2,500 (USD 0.28) per m^3 and Rp. 106,322 (USD 11.81) per m^3, respectively. In Malinau, the average fees received are Rp. 23,040 (USD 2.56) per m^3 and Rp. 23,963 (USD 2.66) per m^3 for monetary and total payments, respectively. For Bulungan, the average fees are Rp. 20,473 (USD 2.27) per m^3 and Rp. 21,269 (USD 2.36) per m^3, respectively; and for Kutai Barat, Rp. 53,863 (USD 5.98) per m^3 and Rp. 54,152 (USD 6.02) per m^3. Thus, communities in Kutai Barat on average received substantially higher levels of fees than those in the other two districts, regardless of whether non-monetary benefits are included or not. The unit of measurement used here is the community and not the contract, as this is the unit used in the analysis in chapter 12. For communities sharing the same agreement, this figure is the same due to difficulties in separating out monetary and some non-monetary benefits from the data and assigning log production to the communities on the contract. Splitting the total benefits and log production evenly between or among communities sharing the same contract results in equal benefits per m^3. As illustrated in figure 9.2, given that relatively few companies complied with any social provisions promised in the contracts, there is very little difference in the values of payments received and payments plus non-monetary provisions, in most cases.

Figure 9.2 Variation in Community Financial and In-kind Benefits from Logging Agreements

i. Monetary payments per m^3

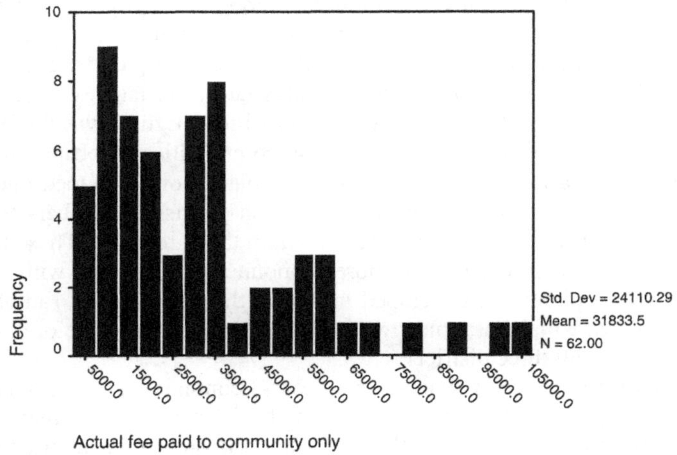

ii. Monetary payments plus value of in-kind benefits per m^3

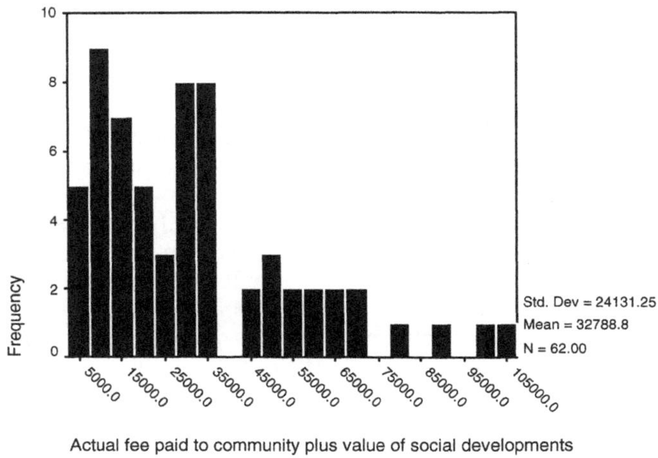

9.4 Firm Compliance and Community-company Conflict

43 communities (69 percent) reported having problems of non-compliance with their respective companies. Many communities reported multiple problems rather than single problems of non-compliance. Problems ranged from late or non-payment of fees to the non-provision of work for local people and a failure to replant logged-over areas. As figure 9.3 demonstrates, the lateness or non-payment of fees (represented by two sets of columns in the figure) was the biggest problem reported by over 50 percent of those communities reporting problems, followed by the failure to reforest or develop plantations (grouped under the heading 'no replanting' in figure 9.3) and the non-provision of other, non-monetary developments and facilities contained within the agreement. These last two categories of non-compliance were more pronounced in Malinau with, for example, not a single plantation developed in any of the communities sampled there. Communities in Kutai Barat meanwhile recorded the highest rate of overall firm compliance of all three districts.

In response to firm non-compliance, 36 out of 43 communities (84 percent) reported to the company first. In Malinau, 14 of the 18 communities that had problems with the company went directly to it, while the other four reported indirectly, through an outside intermediary. In Bulungan, 11 out of 14 communities reported directly to the company, one went to the district government and complained, one reported to the company via an outsider and one did nothing at all. In Kutai Barat, all 11 communities reported directly to the company with a further two also reporting to the district government and another to the local police.

Figure 9.3 Problems with IPPK/HPHH Companies as a Proportion of all Communities within Each District

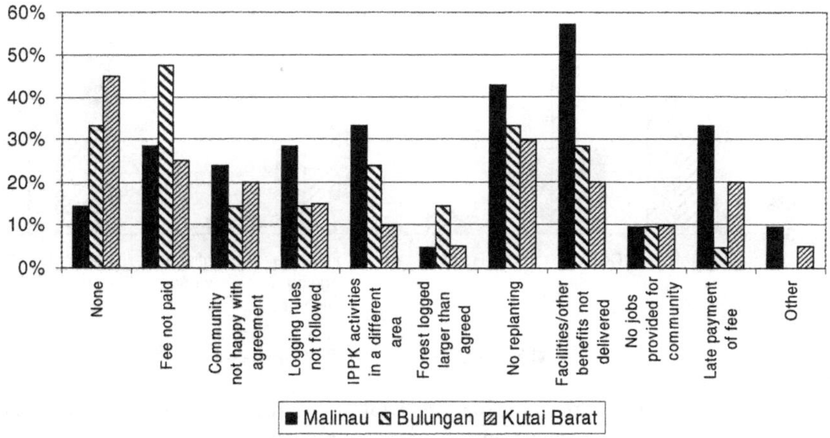

For those communities that had problems with the companies, they were also asked about the role of the district government in the duration of the conflict. In Malinau, all communities responded that the district government played no role whatsoever. In Bulungan, 10 out of 14 communities having problems responded similarly, while three communities said that the local government asked the companies to honour their contracts and in two of these cases actually succeeded in resolving the disputes. In Kutai Barat, of the 11 communities reporting problems, eight responded that the local government had no role whatsoever. In the remaining three cases, the companies were asked to honour their contracts by the local government. In these cases, the local government also helped to resolve the problems eventually despite there being actions undertaken by the communities against the companies as well.

Overall, 17 out of 43 communities (40 percent) with problems of firm non-compliance undertook no further action against the firms, mainly because problems were resolved and there was no need (see figure 9.4). In the remaining cases, communities undertook one or a combination of actions against the companies, with demonstrations and the confiscation of logging equipment being the most common ones. Communities in Kutai Barat that undertook actions were the most likely to stop company operations (in over 70 percent of cases) compared to those undertaken by communities in the other two districts.

Figure 9.4 Proportions of Community-company Disputes resulting in Actions against the Companies

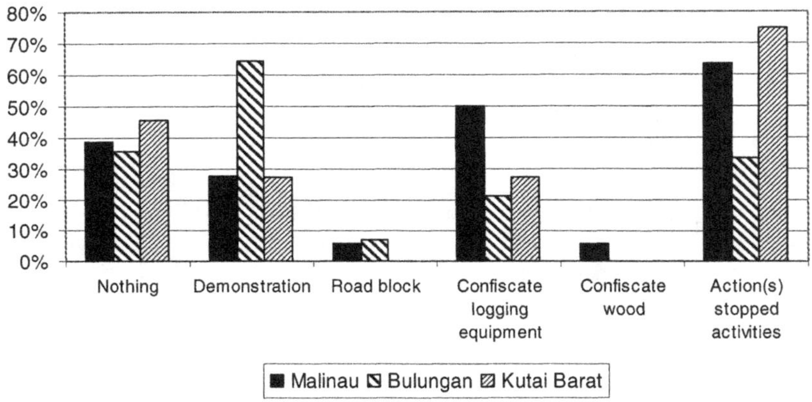

In 18 cases out of 26 communities undertaking actions (69 percent), all groups in the community were involved. In the remaining nine cases, only certain segments of the community were involved in these actions. Household data suggests that overall in those communities where community-firm conflict took place, an average of almost 34 percent of households per community participated.

The maximum length of time that communities participated varied widely although in most cases actions did not last longer than a few days. For example, Sekatak Buji confiscated logging equipment at least twice, but no longer than one or two days at a time, whereas Tanjung Lima, Tajan and Sebatiung undertook a similar action for a month before the company relented and paid its fees.

To settle conflicts, there was mediation between the parties before the company continued logging, in seven out of 26 cases (27 percent). Mediation was typically undertaken by local government officials. In only a single case was a company shut down altogether. In all the remaining cases, the parties settled their differences before logging continued. Sometimes, first agreements were renegotiated during the settlement of disputes. As with the non-conflict situations detailed in chapter 8, most of the changes involved fee payments or trade-offs between fees and social provision such as building materials.

9.5 Conflict with Outsider Groups

Overall, 35 communities (57 percent) indicated problems and conflicts with outsiders as a consequence of their IPPK/HPHH concessions (see figure 9.5). These included conflict over logging borders, company transgressions and disagreements with district governments. Communities in Malinau had a higher rate of these kinds of problems, predominantly with neighbouring communities.

Figure 9.5 Proportions of Communities Having Problems with Outsider Groups since IPPK/HPHH Activities Commenced

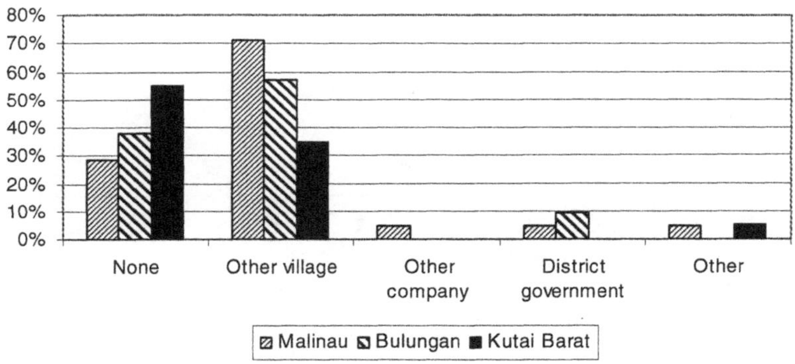

34 out of 62 communities (55 percent) reported having one or more problems with other communities relating to IPPK/HPHH concessions. Of these, 32 reported conflicting forest claims; nine reported logging impacts from other communities' companies; seven reported problems over logging payment; and, seven reported problems with company transgressions over borders as a cause of inter-community conflict. Figure 9.6 gives the causes of inter-community conflict by district.

It seems that a significant number if not all of the problems relating to border problems and company transgressions related to forest claims predated the onset of small concessions negotiated by communities. In a more open political environment in the post-Suharto era, community-company agreements acted as a catalyst for many of these conflicts to come out into the open. For example, logging companies operating for the communities of Setarap and Paking Lama in Malinau both crossed over borders that had been demarcated by Setulang village. Setarap also had border problems with two other communities, Sentaban and Batu Kajang, while Paking Lama had similar problems over borders with Sentaban, Semolon and Long Sulit. In the latter case, until the sub-district heads from three sub-districts intervened and mediated an end to the disputes, both

Setulang and Sentaban demonstrated to the company. Each community blocked the road and stopped company activities for up to two months. This example illustrates the problems that plagued many community-company agreements. That is, overlapping, conflicting border and forest claims, some of which had historical precedent and were only beginning to be addressed for the first time (see for example, Anau et al., 2002).

Figure 9.6 Causes of Problems between Communities as a Proportion of all Communities Sampled within Each District, in Duration of IPPK/HPHH Operations.

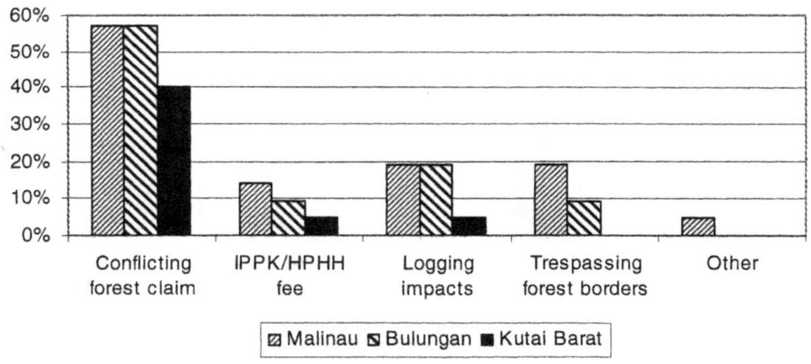

9.6 Environmental Impacts of IPPK/HPHH Operations

Similar to impacts from HPH operations before decentralization detailed in chapter 6, survey respondents were asked about their perceptions of changes (positive, negative, no change) from IPPK/HPHH operations for: river water quality (for washing and drinking); flooding and water levels; hunting; collection of forest products; and forest utilized by the community (for agriculture and gardens). Again, for responses of 'positive' or 'negative', they were asked why they thought there had been a change.

Figure 9.7 gives the average proportion of households per community in each district that perceived these negative impacts only, which were either partially or completely caused by mechanized logging from IPPK/HPHH operations. Thus, a 100 percent response indicates that all households sampled in a particular district had perceived a negative impact say on drinking water quality from IPPK/HPHH logging. The figure also presents the proportion of communities per district giving similar, negative responses (from the focus group meetings). Here, a 100 percent response for a particular district indicates that all sampled communities (on the basis of the focus groups) perceived negative impacts from IPPK/HPHH logging.

Figure 9.7 Community and Household Perceptions of Environmental Impacts from IPPK/HPHH Operations

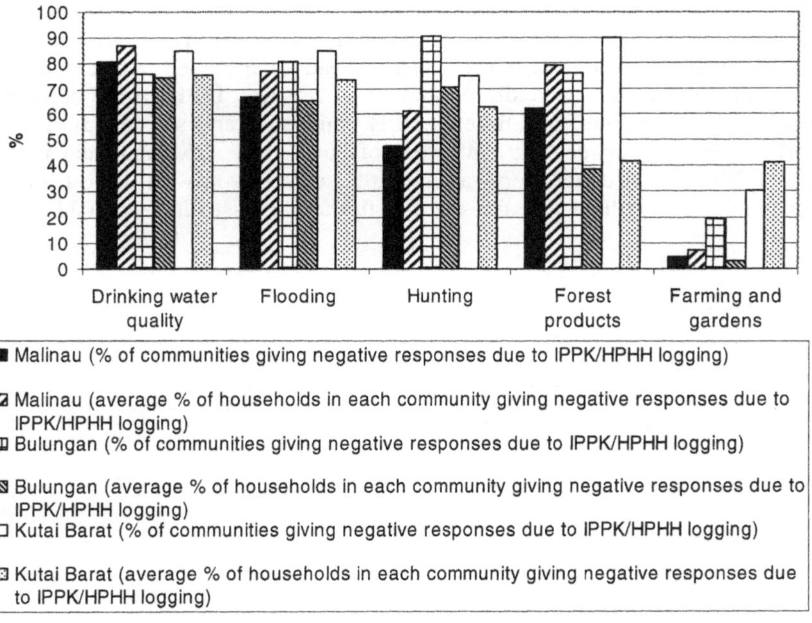

In general, a large majority of communities in all three districts indicated negative impacts on the quality of drinking water in rivers and increased flooding, particularly in Kutai Barat. Moreover, in all three districts, an average of at least 60 percent of households per community gave a similar response for these environmental indicators. Almost 90 percent of communities in Kutai Barat indicated a negative impact on forest products from IPPK/HPHH operations. Just over 40 percent of households on average in Kutai Barat also stated a negative impact on the collection of forest products. For other households, increased and easier access to forest products as a consequence of IPPK/HPHH concessions opening up the forest may have given the impression of 'a positive impact'.

Compared to other indicators, farming and gardens scored the lowest proportion of negative responses from both the community and household surveys in all three districts. Nevertheless, the proportion of negative responses from both the community and household data for farming and gardens in Kutai Barat were on the whole higher than the other two districts. This was maybe because of the intensity and numbers of overlapping concessions in the areas where these communities were surveyed. For example, almost all HPHH areas in Kutai Barat were situated in HPH concession areas, of which almost all had been logged previously. Moreover, in a large number of cases in this district, IPPK/HPHH companies were operating at the same time as HPH companies. Overall, of the 37 cases (60 percent of the sample) claiming that the IPPK/HPHH concession area overlapped with HPH concessions, 28 had been logged before, while the remainder had always been idle. Moreover, in 18 cases, IPPK/HPHH companies were operating at the same time as HPH companies and six communities reported conflicts between the two types of companies. Environmental impacts from IPPK/HPHH concessions, along with a comparison with those from pre-decentralization HPH concessions will be further discussed in chapter 11.

10 Distributional Aspects of Outcomes

Based on the observations about the processes of negotiations, the involvement of brokers described in chapter 7, and the variation in promised fees and production-based payments in chapter 8, this chapter focuses on the distribution of IPPK/HPHH benefits in communities. First, the breakdown of rents flowing from timber buyers and contractors is examined, followed by the results of what happened to the distribution after it flowed to the wider community from the brokers and community leaders.

10.1 The Distribution of Total Rents and the Role of Brokers Revisited

In Malinau and Bulungan, it was usually the brokers who applied for and held permits. These brokers also received separate payments from the timber buyers for their costs in addition to their services, which communities would not have necessarily known about. Brokers were typically given up-front payments for their dealings with the district government and any advance payments to communities. They then received a production-based fee per m^3 that was supposed to cover all arrangements including their services, further payments to the community and further dealings with district government officials. Anecdotal evidence collected in Malinau suggested that this might have been at least Rp. 120,000 (USD 13.33) per m^3 (see figure 10.1). Hence, brokers had a direct incentive to maximise their own benefits by minimising payouts to communities. Moreover, since these brokers were mostly from outside the communities with which they made deals, this also meant that a sizeable proportion of timber rents remained outside the communities.

The level of negotiated fees in Kutai Barat for those communities that owned permits and made agreements with contractors directly was at an average of Rp. 80,789 (USD 8.98) per m^3, lower than the supposed Rp. 120,000 per m^3 received by brokers in Malinau. This suggests that the use of brokers and smuggling routes from Malinau to Malaysia, necessitating more unofficial payoffs along the way, sharply increased transaction costs. The higher prices fetched for timber in Sabah may have offset these costs. Brokers in Kutai Barat tended to receive direct fees from the contractors, separate from that received by the communities, which anecdotal evidence suggests may be in the region of Rp. 5,000 per m^3 (see figure 10.2). This relatively small fee perhaps reflects the diminished role of the broker in Kutai Barat compared to the other two districts. As figure 10.2 illustrates, contractors paid logging fees directly to communities and not via middlemen or brokers, even where they were used.

Figure 10.1 Generalised Structure of Distribution of Negotiated Rent Flows: Malinau and Bulungan

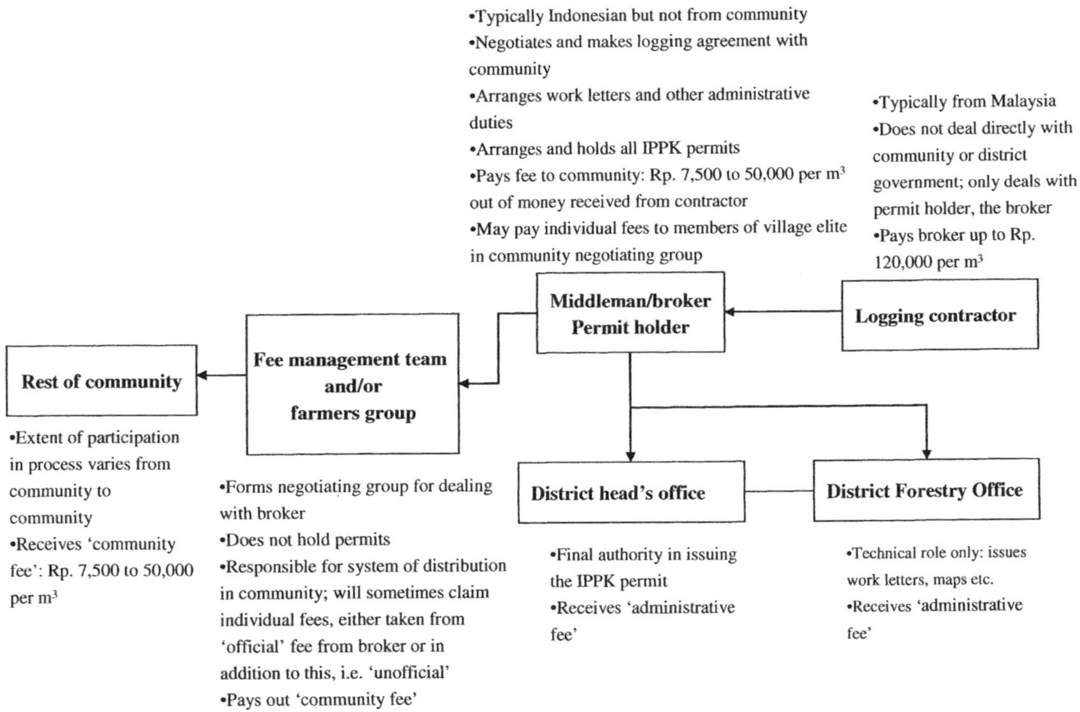

Source: Author's illustration

Notes: Lines with arrows denote direction of log rent payments. Lines (no arrows) denote administrative or institutional links.

Figure 10.2 Generalised Structure of Distribution of Negotiated Rent Flows: Kutai Barat

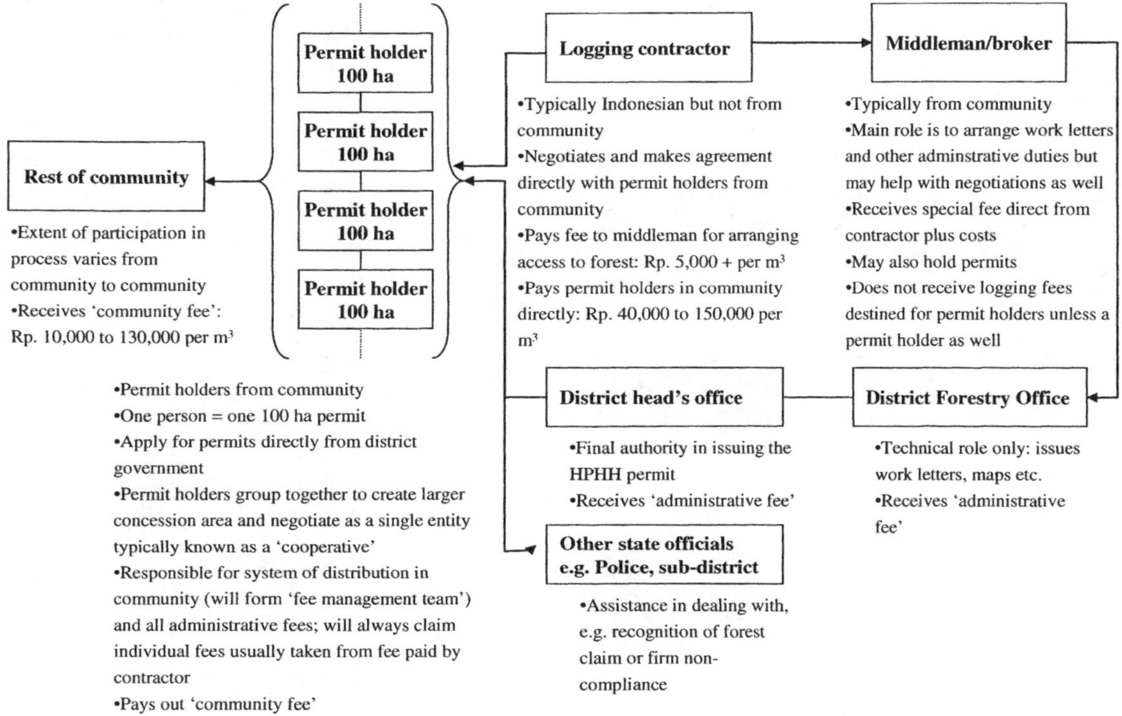

Source: Author's illustration

Notes: Lines with arrows denote direction of log rent payments. Lines (no arrows) denote administrative or institutional links.

Within the community, a defined individual such as the village head or a group of individuals formed from the village leaders or elite received the total fee first before supposedly allocating it the community (see figure 10.3). Overall in the sample, the village head had the sole responsibility for receiving fee money in 29 percent of cases, followed by 'other named select individuals' in 21 percent of cases. These individuals were typically members of elites and village elders.

Figure 10.3 Identity of First Recipients of the IPPK/HPHH Fee Money

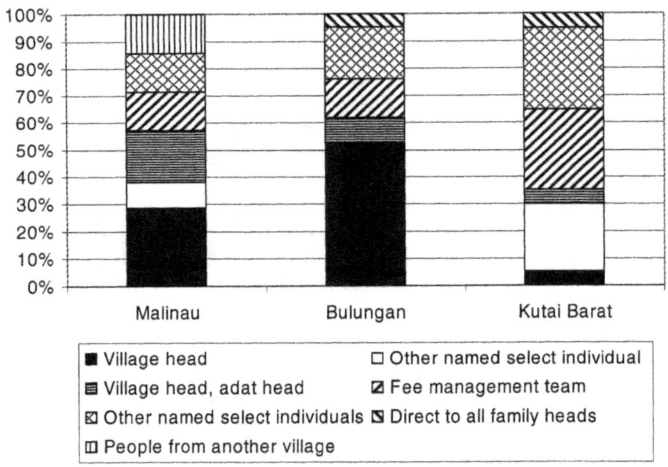

In cases where the agreement was shared by two or more communities, mostly to be found in Malinau, respondents normally gave the name of individuals or group of individuals from one of the communities sharing the agreement. Kutai Barat had the highest number of teams or groups established specifically to manage the HPHH fee money ('fee management teams', usually known as pengurus). These teams did not necessarily contain people who would be considered to be the village leaders or even part of the village elite. They were supposedly selected by community members for this role. There were also two cases (Sekatak Bengara and Temula) where the fee monies were paid directly from the companies to community members in the wider community.

In Kutai Barat, typically only a small proportion of the total fee was set aside to be distributed among members of the wider community ('community fee'). The remainder would be paid out as fees to the permit holders, which were used partly to cover their costs, and fees set aside for development and administration, amongst others. In figure 10.2, the difference between what was paid by the contractors to the permit holders and that received by the rest of the community was the proportion of the overall fee retained by the permit holders. Box 10.1 details the breakdown of the fee paid to Batoq Kelo and a community in the administrative village of Mamahaq Besar, Sungai Mas. Thus, it can be seen that the total fee declared included all personal fees to individuals as well as 'community' fees and even operational costs in the case of Sungai Mas, although the non-labour costs to community employees remained with the contractor.

Box 10.1 Fee Structures for Batoq Kelo and Mamahaq Besar (Sungai mas) in Kutai Barat (all Rp. per m^3)

Batoq Kelo		**Mamahaq Besar (Sungai Mas)**	
Community fee	40,000	Village fee	10,000
Fee for Chairman of management team	20,000	Fee for permit owners ('community')	25,000
Fee for village head	2,500	Fee for development of cooperative and labour	20,000
Fee for members of team (3 people)	15,000	Reforestation fee	17,000
Fee for sub-district head	5,000	Operational HPHH fee	26,000
Fee for police	1,000	Social fee	1,500
For for army	1,000	Education fee (scholarship)	2,000
Fee for sub-district government	1,000	Other fee (*muspika*)	1,000
Fee for sub-district head's secretary	500	Unaccounted for	20,000
Reforestation fee	9,000		
Fee for customary law head	5,000		
Fee for village head's office (admin)	5,000	**Total:**	**150,000**
Fee for customary law head's office (admin)	5,000		
Total:	**110,000**		

Box 10.1 illustrates that the administrative costs of applying for the HPHH permits were borne by these particular communities. There were also a number of other fees that may be termed 'wild taxes', which were paid to all manner of third parties, including the police and army. In fact, it may not be a coincidence that when the company working on behalf of Batoq Kelo failed to make a payment to the community at one point, the community demonstrated against the company and were assisted by the police. This was the only case in the whole sample in which the police enforced the contract on behalf of the community. The question here is whether these payments should be construed as an 'unofficial tax' into the system or a 'bribe' to obtain cooperation.

In the administrative village of Mamahaq Besar where there were three separate negotiating communities all making their own agreements with different companies, the fee structures were even more complex. Within each community, there were 'fee management teams' (*pengurus kelompok*), sometimes numbering no more than two or three members including the chairman. The 'fee management team' also held some of the permits and hence, they claimed separate fees, at least partly to cover the costs of the permit application process. The remainder of this money was paid out to other members of the community who were not members of the 'team'. Moreover, there was a separate 'village fee' (*fee kampung*) that was supposed to be paid to people not belonging to any of the communities in the administrative village, e.g. relative newcomers or migrants who had relatively little claim to the forest claimed for the HPHH concessions. However, the community of Sungai Mas in Mamahaq Besar decided to divert the village fee to a village development fund for the establishment of sewer systems and electricity, instead of allocating it directly to non-member households. One reason for this action was the sudden rise in and squabbling over claims for portions of this 'village fee' among the newer arrivals to the village and older residents not belonging to one of the 'HPHH communities'.

In at least 18 cases (29 percent) fee money was kept for future community developments and not just distributed to community members. There may have been more cases, although details of fee breakdowns are not available for all communities. However, there were also cases in which these funds were sequestered by community members or even distributed among them. For example in Selidung, there was a Rp. 2,000 (USD 0.22) per m^3 'scholarship stipend', which was originally included in the Rp. 30,000 (USD 3.33) per m^3 fee to the community. This was to be saved in a central fund so that children's future school and university costs could be subsidised by the community. Some months later, community members changed their minds about the fund and voted to have this money distributed among themselves instead.

10.2 Community Fees

Households were asked how much was promised to the community as a whole ('total fee') and from this, how much was promised for distribution among the wider community ('community fee'). There were no perceived differences between total fees and community fees in Malinau. In Bulungan, there were three cases with a perceived discrepancy of around Rp. 5,000 (USD 0.56) in each case. However, there is anecdotal evidence which suggests that it was very common for village heads and other members of the elite in Bulungan to personally receive Rp. 5,000 per m^3 or more. These payments were usually passed off as administrative fees[42], although they tended to remain 'undeclared' to the

[42] These types of fees were also observed in numerous cases by Suramenggala et al. (2001).

community at large. By contrast, in Kutai Barat, there were numerous discrepancies between the 'total fee' as perceived by the majority of households in each community and the 'community fee'. There were 12 cases in which the 'community fee' was perceived as lower than the 'total fee'. However, this did not imply that the households in these communities knew about the allocation of the remainder of the total fee (see chapter 13). In the other eight communities in Kutai Barat, it seems that households perceived that the whole of the 'total fee' was distributed to community members, although anecdotal evidence suggests that this was not always the case.

From chapter 8, the average promised 'total fees' net of permit costs in Malinau, Bulungan and Kutai Barat were Rp. 27,125 (USD 3.01), Rp. 32,452 (USD 3.61) and Rp. 73,900 (USD 8.21) per m^3, respectively. But did communities in Kutai Barat really negotiate higher fees than those in Malinau and Bulungan? A proper and direct comparison of promised fees across communities would need to account for all non-monetary benefits. Non-monetary developments were only a feature of around half of the agreements in Kutai Barat (and those with fees at the lower-end of the scale as well), compared with almost all first agreements in Malinau and Bulungan. While there is relatively good data available for promised social developments, the complete lack of proposed production schedules in these agreements makes a production-based valuation and comparison impossible. Instead, a comparison of 'community fees' and agreements not containing any non-monetary provisions follows.

In Malinau, Tanjung Nanga negotiated the highest fee (Rp. 50,000 (USD 5.56) per m^3) in an agreement not including any non-monetary developments. The highest promised fee in Bulungan was Rp. 45,000 (USD 5.00) per m^3, which also included a few non-monetary promises as well (but no private fees to community leaders). In both districts, these promised 'total fees' were equivalent to what households perceived to be the promised 'community fee'. These fee levels in Malinau and Bulungan are, however, only just on a par with the average promised 'community fee' surveyed in Kutai Barat, Rp. 50,900 per m^3, and substantially less than the average 'total fee' there. Therefore, on average, communities in Kutai Barat did appear to negotiate higher fees than those in the other two districts.

10.3 The Distribution of Community Fees

In all surveyed communities, systems of distribution of fee monies were established. It should be noted that this did not imply that the fee money was actually distributed. Overall, systems of fee distribution by 'classes' were established in 40 percent of cases. The remainder simply divided up the fee money equally either per person or per household. Class systems predominated in Malinau and

Bulungan (see figure 10.4), which typically involved three or four classes, based on age, marriage status and whether working or not. For example, in Tanjung Lima, a widow or widower received 75 percent, and a youngster (working but without children to support) received 50 percent of what the head of a household with a family expected to receive.

Figure 10.4 Systems of IPPK/HPHH Fee Distribution

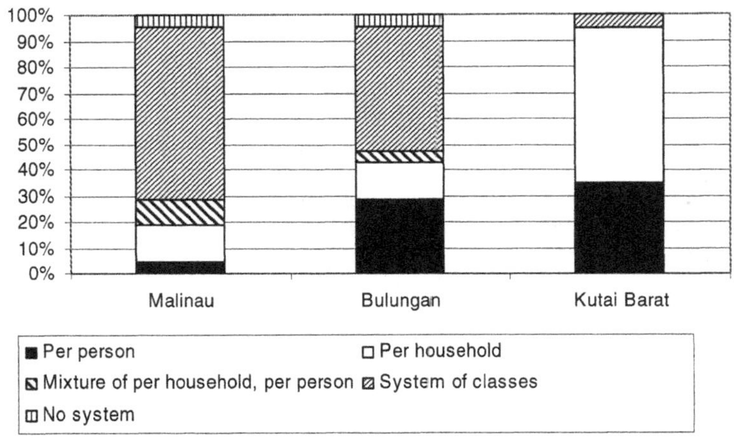

Sometimes, a class system would also have one or more tiers in which people would be further sub-divided on the basis of whether they were relative newcomers to the community or 'original' inhabitants. Batoq Kelo had a system similar to that in Tanjung Lima (three classes: heads of households, unmarried youngsters and divorcees and widows) that was further sub-divided into 'original inhabitants' and 'new inhabitants'. The former group was defined as those people who had lived in the community prior to it moving to its current location in 1983, and accounted for approximately 70 percent of the population in 2003. Alternative systems simply divided the money up on a per person (23 percent of the community sample) or per household basis (29 percent of the sample), and these predominated in Kutai Barat.

Each sampled household was asked how much money they received in total from the IPPK/HPHH concession. Given that this was perhaps one of the most sensitive questions asked in the household survey, the response rate at 95 percent of the total household sample was surprisingly high and similar to the rates for other questions in the survey. On average, each household surveyed in East Kalimantan received Rp. 5.4 million (USD 600.00), with wide variation from district to district. Households in Kutai Barat received the highest average amounts, Rp. 9.4 million (USD 1044.44), followed by Malinau (Rp. 5.5 million or USD 611.11) and Bulungan (Rp. 2.0 million or USD 222.22). Within districts, there was also wide variation among communities. In Bulungan, community averages varied from Rp. 0.2 million (USD 22.2) per household to as high as Rp. 6.9 million (USD 766.67) per household. Community averages in Malinau ranged from Rp. 0.7 million (USD 77.78) per household to Rp. 18.8 million (USD 2088.89) per household, while those in Kutai Barat varied even more, from Rp. 0.7 million (USD 77.78) per household to as high as Rp. 46.2 million (USD 5133.33) per household. There was also wide variation in the levels of payments received among households within many communities, although this appeared to depend on the nature of the fee distribution system established in any particular community.

After asking how much money was received, the household survey asked whether or not the amount of fee money it received was 'fair' given the system of fee distribution, and if not, why not. Despite the subjective nature of the idea of 'fairness', for those who responded "no", their reasoning tended to revolve around complaints about the amounts received not being enough or not as much as other households. Overall, an average of 58.2 percent of households per community thought that the system was 'fair', which breaks down to an average of 64 percent for households in Bulungan, 61 percent in Kutai Barat and 50 percent in Malinau. So while households in Bulungan on average received much less than those in the other districts, this was more likely to be considered 'fair' given the system of distribution, compared to households in either Kutai Barat or Malinau.

10.4 Intra-community Conflict from Rent Distribution

In both surveys, respondents were asked about any problems in their communities resulting from the distribution of fee money in the community. From community level responses, over half of surveyed cases in Malinau and Bulungan, and 75 percent of Kutai Barat communities reported no problems whatsoever. Overall, 61 percent of communities claimed to have had no problems. Figure 10.5 illustrates the nature of the problems that occurred within communities.

While complaints from the community focus groups centred around external claims on the fee money and 'greedy households', household responses tended to focus more on problems relating to distribution within their communities. In particular, a large number of households described perceived differences between what was planned or promised and what was actually distributed to them. Perhaps unsurprisingly, the level of complaints from households easily outstripped the levels recorded at the community level.

Figure 10.5 Problems within Communities as Result of Fee Distribution (Community Data)

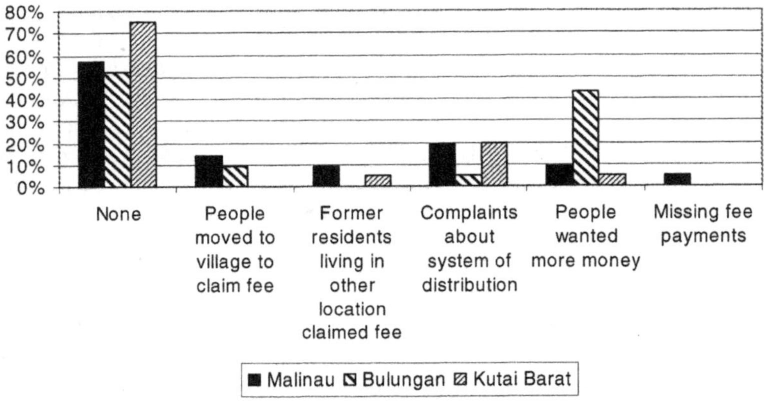

Given the difficulties in comparing problems from one community to another (in terms of seriousness), the variation in nature of complaint plus the difficulties in corroboration, figure 10.6 presents the proportions of households making such complaints about the distribution of fee monies within the community. In only 14 cases (nearly a quarter of the communities sampled) did no households complain at all. In 33 cases (53 percent of the sample), a majority of households (51 percent or over) indicated that there were these kinds of problems in their communities. In only seven cases did all (i.e. 100 per cent) households sampled indicate such problems. By comparison, 19 of the 21 community level responses that indicated problems within their respective communities had household response rates of 51 percent or more. Complaints varied from a general observation of there being problems over the distribution of fee monies, to oblique references to 'problems with the village head' and suspicions towards the people responsible for managing the logging fees. There were many open accusations of corruption and nepotism directed towards village leaders as well.

While there was little evidence of intra-community conflicts becoming violent, there was evidence of deep divisions and a lack of trust among members of some communities. For example, in Bitt Unyang in the administrative village of Ujoh Bilang, a number of community leaders and well-positioned members did not honour their internal fee distribution agreement made with other members of the community. Instead, they kept most of the fee monies when it was received, left the village and relocated to the provincial capital. In this community, over 85 percent of households responded that there were conflicts as a result of this behaviour. Moreover, during the interviews in this community, there were still obvious feelings of anger and bitterness towards those that had left the village, and their families.

Figure 10.6 The Proportions of Communities indicating Intra-community Problems from Logging Fee Monies by Proportion of Sampled Households

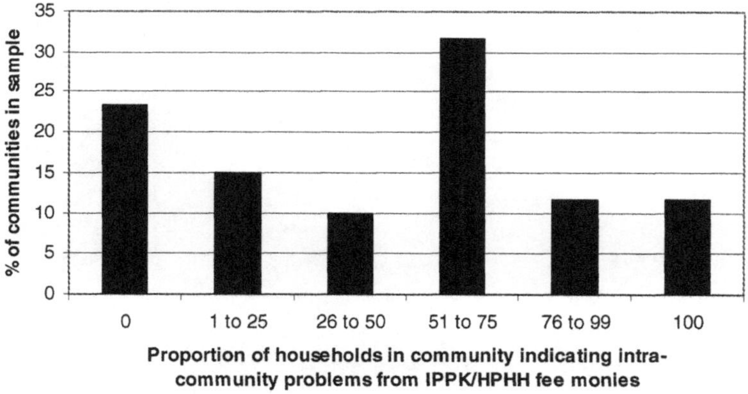

11 Comparison of Impacts of Mechanized Logging on Communities before and after Decentralization

In this chapter, the impacts of decentralization outlined in the conceptual framework described in chapter 3 are analysed for those 60 communities in the sample that experienced mechanised logging both before and after decentralization, followed by an analysis of possible trade-offs among impacts. Considerations of equity and democracy are discussed in chapter 13.

11.1 Monetary and Non-monetary Benefits

As discussed in chapter 6, only eight communities received any kind of compensation or social assistance before decentralization. In table 11.1, data from table 6.1 is presented alongside the IPPK/HPHH benefits received by the same communities. With the exception of community seven, the majority of households in all of these communities claimed to have received monetary benefits from IPPK/HPHH deals. Less than 30 percent of households in community seven received monetary benefits. On average these communities received Rp. 19,455 (USD 2.16) per m^3 (including value of non-monetary developments), a figure which compares unfavourably with the average for the 60 communities in the sample, Rp. 33,055 per m^3 (USD 3.67). So, while the remainder of the sample received little or no benefits from logging operations prior to decentralization, this seems to have had little impact on the levels of benefits received after decentralization. Moreover, those communities that did receive benefits before decentralization did not necessarily appear to have higher expectations or experiences that enabled them to benefit more than those that did not.

The data listed in table 11.1 and similar data for the rest of the sample is analysed to see whether there was any significant statistical difference between financial and non-monetary benefits from mechanised logging received by households before and after decentralization. As described in chapter 6, data for absolute or production-based levels of monetary and non-monetary benefits received by households before decentralization were either unavailable or unreliable. Instead, the differences between the proportions of households within each community claiming to have received financial and non-monetary benefits from companies operating before and after decentralization are analysed. For communities that received nothing before decentralization, these are all zero. The remainder are listed in table 11.1. All communities negotiated for and received some financial benefits, while only 37 out of the sample of 60 communities negotiated non-monetary benefits, not including employment, after decentralization (62 percent).

Table 11.1 Financial and Non-monetary Benefits of Mechanised Logging before and after Decentralization

Comm-unity	Benefits before decentralization			Benefits after decentralization			
	Money (total, IDR million)	*Non-monetary*	*Years of Logg-ing*	*Money (total, IDR mil-lion)*	*Non-monetary*	*Years of Logg-ing*	*Value of benefits[1] (IDR per m^3)*
1	None	Housing (10) 2	30 +	2000 (100)	Generator, transportation to fields (80)	2	33,333
2	None	Farm tools (12,5)	10-15	850 (100)	Housing (100)	2	14,167
3	None	Agricultural assistance, clean water, ground levelling, housing materials (90)	20-25	500 (100)	Scholarship, housing materials, medical assistance, solar panels for generator (50)	2 +	5,000
4	None	Seeds, development of village lavatories (10)	30	480 (89.5)	Village building (10)	3	20,000
5	None	Mosque (7)	25-30	600 (100)	Clean water, generator (14)	2	14,500
6	None	Mosque, agricultural assistance, seeds (37)	30	240 (54.5)	Road, housing materials (20)	2	44,167
7	None	Mosque, agricultural assistance, seeds (57)	30	195 (28.6)	Road, housing materials (29)	1	8,864
8	None	Transportation to fields, agricultural assistance, generator, housing materials (30)	25	989 (100)	Road (60)	2	15,609

Note 1 – Value of financial and non-monetary benefits received by communities divided by log production. IDR denotes Indonesian Rupiah; exchange rate IDR 9,000: USD 1.00; details on benefits derived from combining community and household responses. 2 - The figures given in brackets denote the percentage of households that confirmed that these benefits were received, either individually by households or by the community as a whole.

A lack of meaningful data on community employment both before and after decentralization excludes these benefits from further analysis[43]. Differences in per-

---

[43] While community employment with IPPK/HPHH operations was clearly not extensive due to contractors preferring to bring in their own logging crews, the survey was not always able to capture

ceptions in the provision of non-monetary benefits before and after decentralization are estimated for these 37 communities only.

Household perceptions of financial benefits received from logging companies significantly increased from an average of 1 percent of households per community before decentralization to over 90 percent after decentralization (see appendix 1). Perceptions in the provision of non-monetary benefits also increased significantly (at the 0.05 level), from 11 percent before decentralization to almost 18 percent afterwards. The receipt of any benefits before decentralization in most cases is non-existent. By comparison, and in almost every community, almost all households received some monetary benefits post-decentralization. Also, some received some non-monetary, provision such as wood for houses or clean water barrels or community developments such as churches or roads. These were either provided by the communities with fee money from the companies, or directly from the companies.

11.2 Social Impacts

This section investigates whether or not decentralization experiences and negotiations with logging companies led to feelings of 'community empowerment'. One way of measuring this was to see if these experiences had shifted peoples' perceptions of forest ownership over time. Households were asked 'who owned the forest?' both before decentralization and in the context of the present time. In both questions, respondents could give as many answers as they wanted. The results are in appendix 2 and illustrated in figure 11.1, the former including all possible combinations of answers.

Figure 11.1 shows that community perceptions of forest ownership changed dramatically from the pre- to the post- decentralization era. The numbers of households claiming that the community owned the forest (either alone or in conjunction with another institution or individual) increased from a quarter of the sample before decentralization to over 90 percent after decentralization. The results of the paired samples test (table 4 in appendix) appear to show significant statistical differences for households stating community ownership, whether alone or not. In particular, the mean difference for those stating 'community' alone is significantly different at the 0.05 level. Those saying 'government' or 'state' alone dropped by 35.7 percentage points, from nearly 40 percent before 1997-98 to just three percent of households, in 2003-04. Households stating 'logging companies' in conjunction with the state dropped from 20.1

---

logging employees according to firm. In some communities, people worked locally for HPH firms or even in different areas altogether and the survey results could not always distinguish among these.

percent to zero, and those stating firms alone fell from 11.1 to 0.3 percent. All of these responses were significantly different at the 0.05 level.

Figure 11.1 Perceptions of Forest Ownership Before and After Decentralization

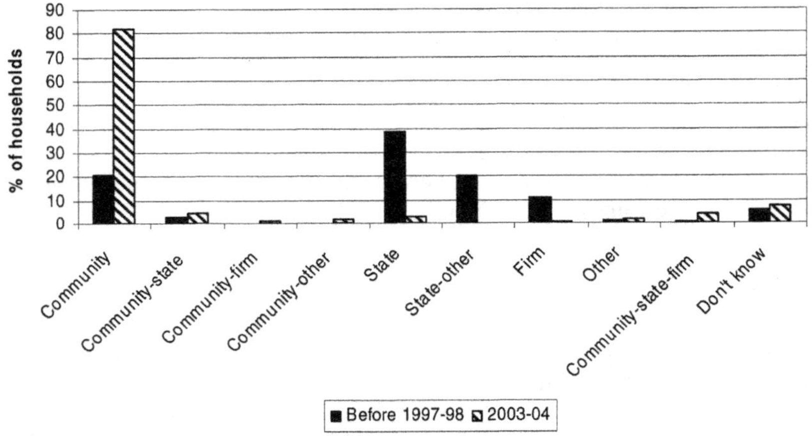

11.3 Enforcement costs

11.3.1 Community-company Conflicts

As described in chapter 6, four of the eight communities that received compensation before 1998 undertook activities against the company beforehand. Of the remaining 52 communities in this sample, 21 or 40 percent of the sample and an average of 18 percent of households per community involved in activities against the HPH companies, ranging from demonstrations to the confiscation of equipment and timber.

After decentralization, a total of 41 out of the sample of 60 communities (68 percent) reported having problems of non-compliance with their respective companies. Again, many communities reported multiple problems rather than a single problem of non-compliance. The trends in the data for these 60 communities follows those illustrated in figure 9.3 for the sample of 62 communities. Overall, the lateness or non-payment of fees was the biggest problem with 53 percent of those communities reporting problems, followed by the failure to replant or develop plantations (37 percent) and the non-provision of other, social developments contained within the agreement (33 percent).

Of these, a further 26 (65 percent of those reporting problems in the community surveys) undertook one or a combination of actions against the companies: 18 undertook demonstrations, 16 confiscated equipment, two blocked roads and two confiscated timber. Where action was actually taken, the communities claimed to have stopped company activities in half of these cases. Also, only 11 out of the 26 cases experienced conflict with logging companies before decentralization as well. In the remaining 15 cases, community members had not participated in activities against companies prior to 1997-98. Of the 25 cases where there had been activities against companies prior to 1997-98, 14 (56 percent) reported no serious problems with their respective IPPK/HPHH companies necessitating direct action. Thus, there seems to be little evidence linking experiences of actions taken before decentralization against companies with those taken in the post-decentralization period. Moreover, contrary to what was expected (see chapter 3), there were almost equal proportions of communities in the sample involved in enforcement actions before and after decentralization. Therefore, there were enforcement costs incurred by communities before decentralization even though these activities were not necessarily successful in claiming compensation from HPH companies.

In 18 cases out of 26 (69 percent), all groups in the community were involved in actions against the company after decentralization. In the remaining nine cases, only certain segments of the community were involved in these actions. Overall, almost 34 percent of households on average per community participated in activities against IPPK/HPHH companies, approximately double the participation rate against HPH companies before decentralization. Possibly greater community empowerment after decentralization and the acknowledgement of community property rights made people less afraid to participate in action against logging companies in defence of their rights. In section 11.4, 'true' enforcement activities are distinguished from 'opportunistic' and 'useless' blockading'.

11.3.2 Inter-community Conflicts

From the community-level surveys, a majority, 33 out of 60 communities (55 percent) reported having one or more problems with other communities. Of these, 30 reported conflicting forest claims (91 percent); nine reported logging impacts from other communities' companies (27 percent); seven reported problems over logging payments (21 percent); and, seven reported problems with company transgressions over borders (21 percent) as a cause of inter-community conflict. A significant number if not all of these problems predated the onset of small concessions negotiated by communities. In a more open political environment in the post-Suharto era, community-company agreements acted as a catalyst for many of these conflicts to come out into the open.

As with the other types of conflict, the relative levels of conflict are difficult to gauge, although again there seems to be little evidence of actual violence between members of different communities. A measure of the seriousness and intensity of these conflicts is derived from household responses. Where problems were less likely to have been resolved and were more serious, a higher rate of recall is expected in those cases compared to communities where the problems were minor or had been resolved already, some time ago. In figure 11.2, the household response rate for the whole sample is included and not just those that were identified by village leaders in the community-level surveys as having involved in inter-community conflicts.

Figure 11.2 Intensity of Inter-community Problems From Logging Fee Monies by Proportion of Sampled Households per Community

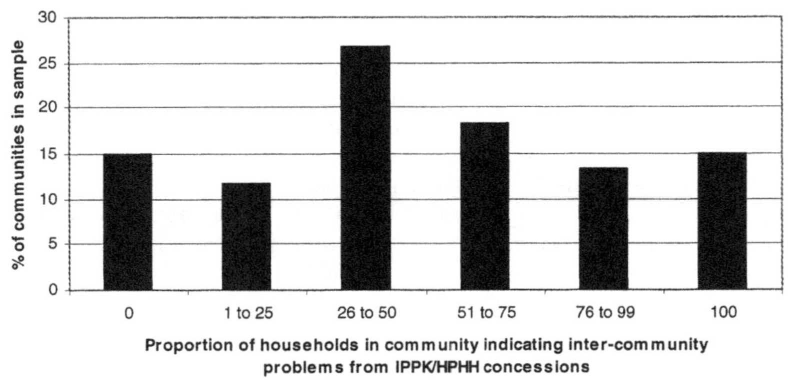

From figure 11.2, a majority of households (51 percent or more) indicated problems with neighbouring communities in 28 communities, or 47 percent of the sample. This compares with the 33 community-level responses indicating similar problems. However, it should be noted that in only nine (or 15 percent of the sample) cases, did zero households indicate that there were no inter-community problems whatsoever; although in another 12 communities only up to 25 percent of households indicated these. This compares favourably with the 27 community-level responses (45 percent of the sample) in which village leaders indicated that there were no inter-community problems as a consequence of IPPK/HPHH agreements.

11.4 Rent-seeking Costs

In 21 cases out of 60 (35 percent), community leaders reported problems in the community as a consequence of the distribution of logging fees. These were typically complaints about how fee monies were distributed within the community, and sometimes even accusations of corruption and negligence by the leaders themselves. Since many leaders were understandably bullish about problems within their own communities, a similar question was asked of households on problems relating to the distribution of rents. The proportions of households making such complaints about the distribution of fee monies for the 60 communities considered here closely follows the trends illustrated in figure 10.6 for the sample of 62 communities.

In only 13 cases (nearly a quarter of the communities sampled) did no households complain at all. In 32 cases (53 percent of the sample), a majority of households (51 percent or over) indicated that there were these kinds of problems in their communities. In only seven cases did all (i.e. 100 percent) households sampled indicate such problems. 19 of the 21 community level responses that indicated problems within their respective communities had household response rates of 51 percent or more. There is no comparable data for the period before decentralization because there were no monetary benefits received by communities and hence, no need for systems of distribution at that time.

There are difficulties in comparing problems among communities in terms of seriousness, the variation in nature of complaint plus the difficulties in corroboration. Nevertheless, it could be inferred from figure 10.6 that the fewer people who responded to this question in the affirmative, the less likely those problems were serious or had long-lasting impacts, i.e., people forgot or resolved them and moved on. Conversely, the higher the proportion of people responding positively to this question, the greater the amount of 'noise' relating to internal problems, and perhaps the greater the severity of the problem.

11.4.1 'True' Self-enforcement', 'Opportunistic' or 'Useless' Blockading?

As discussed in chapter 3, there is a need to empirically distinguish 'true enforcement' activities undertaken by communities against companies from 'opportunistic' and 'useless' blockading. A 'true enforcement' activity is a response by communities to non-compliance by firms, while an 'opportunistic' activity is one undertaken by community members in order to seek rents for themselves and when the company has at least partly complied with the agreement made with the community. 'Useless' blockading on the other hand may also occur when the company is in compliance, although in this case, rent-seeking community leaders prefer to keep the benefits for themselves and blame the non-distribution of these to the wider community on company 'non-compliance'.

Data for intra-community problems can be combined with data for firm compliance and participation in community-firm conflicts to help identify 'true' enforcement actions from those driven by rent-seeking behaviour. Higher household participation rates may be closer to 'true' collective actions, while 'useless blockading' may be identified by firm compliance and higher intensities of intra-community conflict. A lower participation rate but above zero could either indicate 'opportunistic blockading' or simply a lack of incentive on the part of community members due to any kind of village elite rent seeking. If the former, then this could be identified by firm compliance too. Figure 11.3 illustrates the outcomes for community participation in community-firm conflicts and level of intra-community conflicts (using household data) with communities marked according to whether the firms had complied with agreements or not (taken from community data). Non-compliance is a discrete variable that is defined as any term in the contract that community leaders had explicitly noted as being a 'broken promise', i.e. includes fee payments, logging rules, employment etc.

Figure 11.3 indicates limited evidence for a trade-off between the intensity of intra-community conflict and household participation in community-firm conflict. Overall, there were 26 cases of firm non-compliance as claimed by community leaders leading to some household participation in community enforcement actions, plus a single case of firm compliance with some household participation in actions against the company. Six of these non-compliance cases in the upper-left quadrant appear to have a high participation rate in community-firm conflicts but a low level of reporting of problems within the community; these communities incurred self-enforcement costs and had fewer rent-seeking problems. These cases may represent the only examples of self-enforcement with 'true' community collective actions. Two cases of firm non-compliance in the lower-left quadrant had lower rates of household participation in self-enforcement, in addition to low levels of intra-community problems. Lower household participation in these cases may have been due to relatively high opportunity or transaction costs, or weak social capital, although they could still be considered 'true enforcement', and not 'useless' or 'opportunistic'. These eight cases, spread out among all three districts, received an average fee of Rp. 43,322 (USD 4.81) per m^3, including the value of non-monetary developments. In the same quadrant, there are three cases of firm non-compliance with zero household participation, which was either due to there being no conflict between the parties or a low community ability to organise. The average payoff for these communities was Rp. 29,289 (USD 3.25) per m^3.

Figure 11.3 Comparison of Household Participation in Community-firm Conflicts against the Proportions of Households reporting on Intra-community Conflicts as a Result of IPPK/HPHH Fee

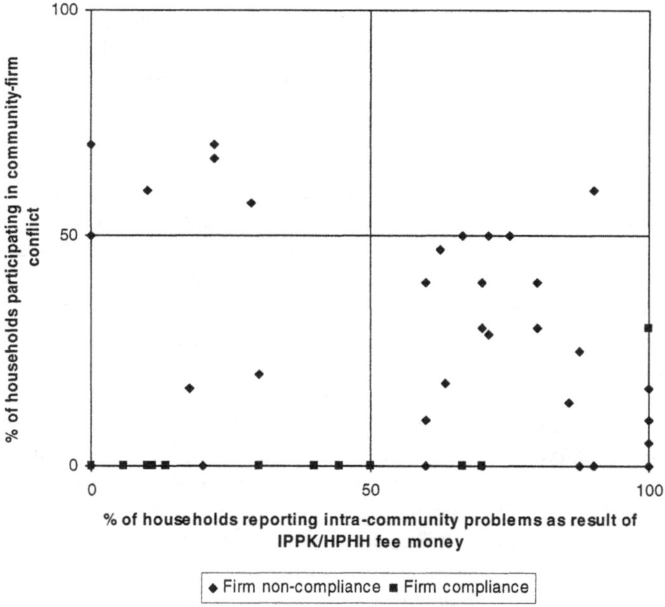

Note Some data points overlap and are not visible on this figure.

Also in the bottom-left quadrant, there are 17 cases of company compliance (eight in Kutai Barat, seven in Bulungan and two in Malinau) with zero participation in community-firm conflicts and relatively low rates of intra-community conflict. With an average payoff of Rp. 42,842 (USD 4.76) per m^3 these cases had the most positive outcomes, with no enforcement costs and little rent-seeking costs incurred.

In the lower-right quadrant, there are three cases of firm compliance but with higher rates of intra-community conflict. Here, high levels of (declared) rent indicates that there might have been rent-seeking by village elites in the two cases with zero participation in community-company conflicts. For these two, the average fee received was Rp. 50,722 (USD 5.64) per m^3 (including the value of non-monetary developments). The lack of household participation implies that no blame was apportioned to the company for non-distribution of rents by the village elites. The third case of firm compliance might have involved opportunistic blockading given a low but positive rate of household participation in community-company conflicts. This community received Rp. 17,500 (USD 1.94) per m^3.

The lower-right quadrant also contains the largest number of communities where firm non-compliance occurred. First, there were eight cases with zero household participation in self-enforcement actions, which was due to there being no actual conflict between the parties. With an average fee of Rp. 28,031 (USD 3.11) per m^3, the relatively high intensities of intra-community conflict indicates that rent-seeking by elites probably occurred, with rent not appearing in fees received.

This quadrant also contains 20 communities with firm non-compliance but with relatively low (50 percent or less) proportions of households actually participating. This includes 18 out of the total of 26 (65 percent) that claimed to have had conflict with their respective firms. Thus, at least two of these cases were not even claimed by community leaders as a community self-enforcement action, and were more likely to be opportunistic. All these cases had higher intensities of intra-community conflict, indicating a higher prevalence of rent-seeking behaviour. The lower rates of household participation for the other 18 cases might indicate opportunistic blockading, followed by firm non-compliance, or rent-seeking by elites (via 'special fees'). Regardless, the household incentive to subsequently deal with firm non-compliance was almost certainly undermined by rent-seeking, which is reflected in the relatively low, average fee received by these 20 communities, Rp. 24,405 (USD 2.71) per m^3. The higher household participation (towards 50 percent of sampled households and including the single case in the upper-right quadrant, which received IDR 24,850 (USD 2.76) per m^3), the greater the likelihood that 'useless blockading' took place before members realised that the elites had captured and not distributed rents within the communities. Again, any actual, subsequent non-compliance by firms is likely to have been met by 'true enforcement' actions that were undermined by rent seeking in the community. However, this kind of analysis cannot distinguish among the exact motivations for actions in the right-hand side of figure 11.3.

The incorporation of other variables into the analysis could be used to help identify the motivations for community-firm conflicts. Opportunistic blockading for example might be identified not only by having relatively lower household participation rates but also lower duration and higher frequency. In order to test the presence of multi-variable clusters in the sample, principal component or factor analysis is used. The idea is to identify underlying variables, or factors that explain the pattern of correlations within a set of observed variables. In addition to undertaking this analysis using the discrete variable for firm compliance, a more sensitive indicator of firm compliance is used: the ratio of actual fees received and promised fee per m^3 of production. Although communities reported on a range of issues in which firms failed to comply, the initial observation that community-firm conflicts were typically settled with cash compensation implies that the fee ratio could be used as a continuous proxy for non-compliance with non-fee related terms as well, with a maximum value of one. In addition to the intensity of intra-community conflict and household participation in community-firm conflict, two other closely related variables are included in the analysis: the duration of the longest conflict with the firm and the frequency of community-firm conflicts that took place in each community. The objective of the analysis is to test for the presence of multiple clusters of observations in a multi-variable framework, thus building on the simple three variable framework presented in figure 11.3.

First, using the discrete variable for firm non-compliance (FIRMNONC), the correlation matrix seems to indicate significance at the 0.05 level for almost all combinations of factors (see appendix 3). The exceptions are intensity of intra-community conflict (INTRACOM) with both duration of the longest community-firm conflict (MAXCONFC) and household participation in community-firm conflict (HHPARTCON). The determinant for the correlation is 0.178 suggesting that it can be inverted. Thus, the greater the likelihood of firm non-compliance, the more intense intra-community conflict, the higher both the frequency and duration of community-firm conflict, and the higher household participation. Also, an increase in intensity of intra-community conflict seems to be correlated with a higher frequency of community-firm conflict. This suggests that there may be a greater likelihood of opportunistic blockading where there are more intra-community problems. Frequency also seems to be strongly positively correlated with duration and household participation. This seems to counteract the idea that higher frequencies of blockading might be due to smaller numbers of opportunistic blockaders. Also, this counteracts the idea that opportunistic blockades might be short-lived. Duration and participation also seem to be strongly, positively correlated, suggesting that longer enforcement activities against firms generally have higher participation from the community. This seems to support the idea that longer enforcement actions require higher community participation and hence, greater collective action to be sustained. Further tests indicate that all data used here is suitable for use in the principle component analysis[44].

[44] The Kaiser-Meyer-Olkin (KMO) measure of sampling adequacy indicates the proportion of variance in the variables which is common variance, and has a relatively high value of 0.725. Bartlett's test of sphericity indicates whether or not there are significant relationships among the variables, which seems to be the case ($p = 0.000$). A study of the anti-image matrices gives an indication of correlations which are not due to the common factors. Only a few correlations have small

Replacing the discrete variable for compliance with a continuous variable based on fee ratios (FEECOMP2) leads to a weaker set of results for the factor analysis (see appendix 4). There are a higher number of insignificant correlations (at the 0.05 level) than in the previous analysis, including fee compliance with HHPARCON, INTRACOM and MAXCONFC. While the results from the KMO and Bartlett's test both still suggest that these data are still suitable for this kind of analysis, the relatively low MSA for the anti-image correlation for FEECOMP2 (0.296) seems to indicate that it does not fit with the structure of other variables. Moreover, the extraction communality for INTRACOMM has declined to 0.427, a relatively low amount of variance that is accounted for by the components in the factor solution compared to before, suggesting that INTRACOMM be dropped from the analysis. Overall, the continuous variable for firm compliance results in a weaker model than when using the discrete variable.

Where the firm has been perceived as complying with all provisions in the agreement, actual fee compliance as a ratio of actual fee received and the promised fee per m^3 of production is higher. However, in many cases it seems that even where community level responses have stated that the firm has not complied (x axis), at least originally, the firm has at some point paid some form of financial compensation for this non-compliance. The discrete measure of firm compliance covers all firm behaviour even when they eventually complied, whereas the continuous measure is the eventual outcome, which might explain why there is not a strong correlation between the two measures. In other words, the eventual outcome is *ex-post*, while enforcement may have responded to *ex-ante* non-compliance.

Another weakness with using the fee ratio measure of actual over negotiated payments is that of disappearing payments due to rent seeking. Thus some of the fee payments received would not show up in the data for actual payments and hence the fee ratio. Another problem is that as described in chapter 9, fees were

values, indicating that they are relatively free of unexplained correlations. The Measure of Sampling Adequacy (MSA) for each factor in the anti-image correlation matrix all have values over 0.5, which indicates that all the variables seem to fit with the structure of the other variables in the model. Extraction communalities are estimates of the variance in each variable accounted for by the components in the factor solution, which are all relatively high, in particular intensity of intra-community conflicts (0.857) and household participation in community-firm conflicts (0.806). The results of the component matrix seems to indicate that firm non-compliance (0.761), incidence of community-firm conflict (0.852), duration of longest community-firm conflict (0.712) and household participation in community-firm conflict (0.844) are all highly correlated with unrotated factor 1 and not factor 2. The intensity of intra-community conflict on the other hand is more correlated with unrotated factor 2 (0.843) than with 1 (0.382). The predicted pattern of relationships seems to confirm that the factor analysis solution is correct. While most of the residuals are very small, there are still seven nonredundant residuals with absolute values greater than 0.05.

negotiated, both upwards and downwards, in quite a number of cases meaning that using 'first promised fees' to derive the fee ratio might skew the ratios one way or another. Finally, it is not always clear that cash compensation was used to settle disputes, suggesting that fee ratio may not be an appropriate proxy for a general measure of firm compliance. Hence, for a more accurate record of firm behaviour in the duration of the actual agreement, the discrete variable is the better one to use in the context of this particular analysis.

11.5 Environmental Costs

11.5.1 Changes in Environmental Impacts from Decentralization

As described in chapters 6 and 9, household respondents were asked about their perceptions of changes (positive, negative, no change) for five selected environmental indicators: river water quality (for washing and drinking); flooding and water levels; hunting; collection of forest products, and; forest utilized by community (for agriculture and gardens). Then, for responses of 'positive' or 'negative', they were asked why they thought there had been a change. The same questions were asked in relation to HPH operations prior to decentralization, and for IPPK/HPHH operations, post-decentralization. Figure 11.4 compares the average proportion of households per community that perceived negative impacts only that were partially or completely caused by mechanized logging, first, for HPH before decentralization, and then, for IPPK/HPHH.

From figure 11.4, an average of almost 80 percent of respondents indicated that water quality had been adversely affected by logging both before and after decentralization. Increased flooding was indicated by over 75 percent of respondents before decentralization, and 70 percent afterwards. Thus, there is very little difference for impacts on water quality and a slightly higher proportion of households responding that impacts from HPH on flood levels were negative before decentralization compared with similar impacts from IPPK/HPHH activities. Over 60 percent indicated that hunting had been adversely affected both before and after. Appendix 5 contains the results for the testing of any statistical differences between local environmental impacts as perceived by communities. Here, for all indicators, each pair is the proportion of households in each community that perceived the specified environmental impacts before and after decentralization. The results show that there are no statistically significant differences between HPH and IPPK/HPHH impacts on water quality, flooding and hunting. Therefore, perceptions of these environmental impacts remain unchanged with decentralization.

Figure 11.4 Household Perceptions of Negative Impacts from Mechanised Logging Before and After Decentralization

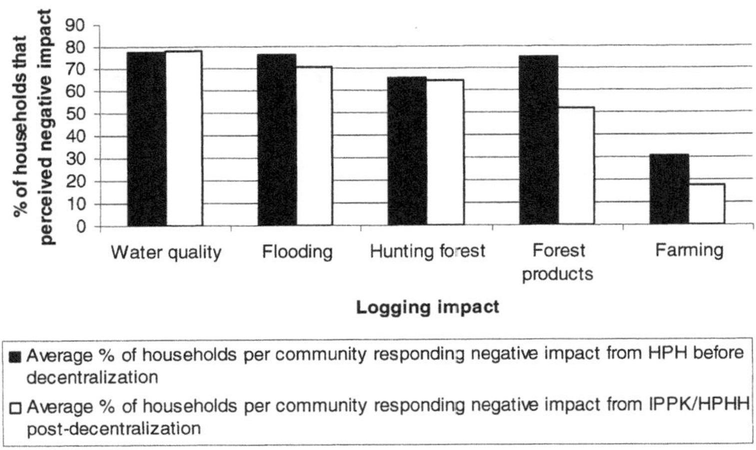

With respect to the collection of forest products, an average of approximately 75 percent of respondents per community indicated negative impacts from logging before decentralization, falling to 50 percent afterwards; the mean difference of 23 percent is statistically significant (see appendix 5). The average perceived negative impacts of logging on forest used for farming and gardens also significantly declines with decentralization, from 30 to 17 percent. HPH concessionaires tended to prevent communities from entering concession areas and utilizing the forest. Thus on average, households were more likely to state that HPH operations had a negative impact on farming and the collection of forest products compared with IPPK/HPHH activities. Selective logging opens up forest, which makes cultivation of the land easier and enables easier access to some forest products, including timber. So, some households responded that IPPK/HPHH activities actually improved farming and gardening for these reasons. These responses are not reflected in figure 11.4, although an average of 8.8 and 9.4 percent of households per community responded that IPPK activities had a positive impact on farming and the collection of forest products, respectively. It should be noted, however, that some of the effects of unimpeded access may not be sustainable and were simply the unintended side-effect of logging.

With the exception of farming and forest products, environmental impacts did not appear to have become better or worse from the perspective of households. However, the relative severity of impacts cannot be inferred from this data nor can it be compared from one community to the next. The data only reveals perceptions of changes. At least 36 IPPK/HPHH concessions, or 60 per cent of the sample were either partially or fully located in old HPH areas. Of these, 28 concessions had been logged previously, which introduces the element of impacts from 'logging on logging'. In this case, more people could be expected to respond in the negative when questioned about impacts and hence, resulting in data showing clearer differences between all post and pre-decentralization environmental impacts than has been shown. There are three possible explanations for these results. First, communities have been known to deliberately underestimate logging damages from IPPK/HPHH operations because they think that by providing support for the system, it could continue without interference from the government (see Limberg, 2004). Second, a number of communities were all located in similar areas, and it was very difficult for some respondents to be able to differentiate among the impacts from a proliferation of concessions. A third possible reason is that communities may have been able to negotiate some environmental controls in the agreements with their respective companies in order to limit local impacts from logging activities.

11.5.2 Trade-offs between the Environment and Other Contractual Provisions

Did communities make (consciously or sub-consciously) or were they forced to make any trade-offs between financial and environmental provisions when negotiating agreements? 39 out of 60 (64 per cent) communities claimed environmental provisions in their HPHH/IPPK contracts. 35 out of the 55 surveyed contracts (also 64 percent) contained some kind of environmental provision. As described in chapter 8, the most common provisions included reforestation, a limit on the diameter of trees to be cut, the species to be removed and the exact location of the concession area. Provisions for non-timber plantations such as rattan gardens, coconut and palm oil plantations are classified here as 'agricultural developments'. These are to be distinguished from 'reforestation', an activity that was supposed to be undertaken by companies, and hence is defined here as an 'environmental rule'.

Table 11.2 compares the contracts with and without any environmental provisions. To test whether there were statistically significant differences between the two types of contracts, the independent samples t-test is used (in table 11.2 * denotes significance at 0.10 and ** denotes significance at 0.05 level). For first promised fee, a high value for the t-test ($p = 0.218$), implies that there is no significant difference between the two group means. The results are similar for renegotiated fees, promises of social developments, promises of employment and promises of agricultural developments as well. In other words, there are no significant differences between the contracts containing any environmental provisions and those containing none at all. Therefore, there is no statistical evidence of any negotiation trade-offs between the environment and other provisions within the agreements in this sample.

For actual fee payments, the t-test indicates that the difference between the two group means is significant (p = 0.004). With respect to the proportions of communities experiencing compliance with social developments, job provision and agricultural developments, the high values for the t-test indicate that the differences between the group means for each of these is not significant. Hence, as before, there is no statistical evidence of trade-offs in actual contract provisions as a result of environmental provisions in community-firm agreements except that actual payments are slightly higher for contracts containing environmental provisions.

In summary, while there is no evidence of a trade-off between promised fees and environmental provisions, there is some evidence that the contracts containing environmental provisions were more likely to have higher actual financial pay-offs compared with those that did not. This suggests communities that negotiated contracts containing environmental provisions had a greater capacity to self-enforce agreements than those with no provisions in their contracts. It is possible that community bargaining power and self-enforcement capacity are linked as shown theoretically in chapter 3. The overall correlation coefficient between promised and actual fees is very high at 0.455), and hence, highly significant.

There is little evidence of significant differences between perceptions of environmental impacts resulting from contracts containing environmental provisions and those that did not. Table 11.2 indicates that the differences between the group means for water quality, flooding and forest products for contracts containing environmental provisions and those containing no provisions are not statistically significant. The differences for negative impacts on hunting and farming are statistically significant at the 0.10 level, although actual impacts on farming appear to be worse for those agreements containing environmental provisions in contrast to all other environmental indicators. Possibly, the inclusion of environmental provisions in the contract also reduced access for community members. Overall, since most environmental indicators are not significantly different, communities that negotiated environmental provisions in their contracts were definitely not worse-off than those that did not negotiate these. Moreover, they may in fact have been better-off in terms of monetary benefits, which indicates that there were no trade-offs between these provisions.

Table 11.2 Comparison of Contracts with and without Environmental Provisions

	Contracts containing :	
	Environmental provision(s)	**No environmental provision(s)**
Number of contracts	35	20
Average first fee promised (IDR per cubic meter) [1]	47,897	38,762
Average fee including renegotiated fee (IDR per cubic meter) [1]	44,615	42,357
% of contracts containing social developments	79.0	71.4
% of contracts containing jobs for community members	87.2	76.2
% of contracts containing agricultural development[2]	34.2	42.9
Average actual payment (IDR per cubic meter)	38,698	22,300**
Average actual payment as % of first fee (as % of final, renegotiated fee): fee compliance level.	80.8 (86.7)	57.5 (52.6)
% of social development promises complied with[2]	76.7	86.67
% of employment terms complied with[3]	85.3	87.5
% of agricultural promises complied with[3]	15.4	11.1
% of environmental promises complied with[3,4]	53.8	-
Average % of households indicating negative impact on water quality[5]	76.4	82.2
Average % of households indicating negative impact on flooding[5]	67.2	77.7
Average % of households indicating negative impact on hunting[5]	60.1	72.3*
Average % of households indicating negative impact on forest products[5]	48.1	58.7
Average % of households indicating negative impact on farming[5]	20.9	10.7*

Note For all fees use conversion rate IDR 9,000 = USD 1.00. 1 - First and renegotiated fees considered if applicable. 2 - Unfortunately a lack of data for agricultural developments meant that it was not possible to include a valuation of these benefits, and here they have been treated as a discrete variable. 3 - Includes contracts only partially complied with. 4 – 'Environmental compliance' is measured only from community level responses about company non-compliance in general. None of the responses were verified by checking concession areas. Any replanting fees are accounted for. 5 – Per contract. For shared agreements, these include all sampled households for communities sharing. *difference between contracts containing environmental provisions and those not containing provisions is significant at 0.10 level; ** difference is significant at 0.05 level.

12 Determinants of Community Benefits from Post-decentralization Logging Negotiations

In this chapter, the hypotheses derived in chapter 3 to explain the variation in outcomes are tested using the data described in Part II. First, the model is operationalized with proxies developed from the empirical data before a presentation of the main econometric results is given. Results are presented for the three hypotheses and a comparison made at the end of the chapter.

12.1 Operationalizing the Model

To operationalize the model described in chapter 3 and to test the hypotheses derived, a number of proxies are developed from the empirical data described in Part II. Most of the proxies were derived from data collected at the community level, while the remainder was calculated from responses given at the household level. These are listed in appendix 6.

For the incorporation of an assumption of a minimum payment by the firm, the model is operationalized in two steps. In the first stage of the analysis, data for all 62 communities is included and the probability of a community being able to obtain a payment above the minimal level is used as a dependent variable (hypothesis set 1). In the second stage, only the sub-sample of communities that are able to obtain a payment above the minimal level are used, in order to explain the variation in their actual payoffs (hypothesis set 2). Since the minimal level is not known with certainty, the lowest fee negotiated or renegotiated in the sample is used as the threshold, which amounted to Rp. 15,000 (USD 1.67) per m^3. To test for robustness, the threshold level is varied. The lack of log production schedules in most contracts means that a threshold including the value of promised non-monetary benefits could not be calculated with any certainty. Thus, hypotheses sets 1 and 2 are tested using data for fee payments alone. Finally, for testing the total effects of the determinants on outcomes, data is included for all 62 communities (hypothesis set 3). Two measures are used for the dependent variable: financial payoff per m^3 and financial plus value of social developments per m^3. These can be seen in chapter 9. The proxies described in this section are used to test the hypotheses both for the effects of parameters on outcomes in individual stages and the total effects of parameters on outcomes.

The community's valuation of the standing forest (*b*) refers to local use values only, and is likely to depend on whether the community expects to stay in the area in the future (proxied by number of years spent in the area and whether it is predominantly semi-nomadic), and the degree to which it competes with neighboring communities for the same forest (number of neighbouring communities sharing a forest border). Moreover, it is likely to depend on the scarcity of the forest (proxied by per capita forest claim), as well as previous experiences with logging and its environmental impacts. Also, it is expected that the higher the proportion of household income derived from forest products and the higher timber quality, the higher the standing value to the community. Forest products in this context does not refer to timber harvested by the firm, but could refer to timber harvested and sold 'illegally' by the household.

NGO activity in providing advice and information on logging impacts is likely to increase *b*. Such activity is, however, likely to be endogenous since NGOs often target communities that are expected to be more disadvantaged in logging negotiations. As a consequence, this proxy cannot be used in the regression without further instrumentation, which was not feasible given the limited number of observations in our sample. Moreover, the value of the forest reflected in actual negotiations will depend on who in the community negotiates on behalf of whom. It is hypothesized that communities in which knowledge of logging agreements is shared more widely will have a higher value of *b* represented in negotiations than those where only a few local leaders know about the process. This proxy is less relevant in determining the community's ability to enforce property rights since it would require collective action by all or many community members.

The profitability of logging is likely to depend on the total area logged, as there may be economies or diseconomies of scale. Moreover, timber from the districts of Malinau and Bulungan tends to be exported to Malaysia where timber prices have been significantly higher than in the Indonesian market where, as detailed in chapter 2, most of the timber from Kutai Barat is sold.[45] Most timber from these concessions was transported by river. While profitability is expected to be decreasing in the distance from timber markets, a lack of data regarding the exact locations of concession areas and timber markets make this difficult to quantify. Instead the district may serve as a proxy for timber markets with Kutai Barat used as an, admittedly rough, discrete variable. The use of brokers also affects timber values, particularly in the case of negotiations, as brokers receive a significant share of the profits. Similar to NGO assistance, this proxy could not be used in the regression without further instrumentation due to the possibility that use of brokers is endogenous. A final proxy for logging profitability is 'forest quality', derived from whether or not the concession area had been commercially logged before negotiations. Here, profitability is expected to be decreasing with previous logging. While, this is a rather coarse proxy for forest quality, using say forest cover data is not very reliable or accurate either because of the tenuous links between deforestation and logging. This is particu-

[45] Of course, higher prices may be offset by the risk of being caught and possibly punished, although there is a relatively low risk of this occurring in Indonesia. The expected value of punishment for perpetuating a forest crime in Indonesia is close to zero (Repetto and Gillis 1988; McGrath and Grandalski, 2000).

larly the case where logging tends to be selective as in East Kalimantan, and where there is a proliferation of overlapping concessions, both old and new.

Timber profitability (and also community valuation of the standing forest) may depend on tree species as well. While very few interview respondents knew about the species harvested from their respective concessions, the limited data collected confirmed the general observation that log production in the area is dominated by the Meranti species. Due to being relatively close together, the surveyed areas also had similar geography, flora and fauna and exposure to commercial logging in the past. Thus, given the unavailability of more precise data, it was assumed that unobserved conditions are roughly identical across the communities surveyed. This also holds for firm-specific data, the collection of which was beyond the scope of this study. That there were no significant differences across firms in operational costs was confirmed by the relative lack of variation in promised fees to communities from say the districts of Malinau to Bulungan, at least compared with the variation in actual payments received.

Proxies for community's blockading costs are largely based on the theory of collective action, including number of households, participation in organizations, and social heterogeneity.[46] Moreover, blockading costs are likely to be increasing in the opportunity cost of time. Off-farm employment in East Kalimantan is mostly in the form of government jobs[47]. By contrast, being located close to the market is likely to reduce opportunity costs as it reduces the time spent in transporting and selling produce, an activity conducted fairly equally by all households involved in subsistence farming. Alternatively, one may consider the more conventional hypothesis that the closer the community is to the market, the higher the opportunity costs of time, thus negatively impacting on a community's ability for collective action.

No proxies for the firm's discount rate and fixed cost of logging are included as data are unavailable. However, these are not likely to vary much across firms in East Kalimantan[48]. Small concessions tended to use a basic minimum of capital equipment in the form of trucks and bulldozers owned by contractors. They also built relatively few roads compared to HPH companies, and as previously mentioned tended to ship their logs downriver (see Limberg, 2004). At the community level, the availability of different types of infrastructure is used as proxy for the community's discount rate, with poorer communi-

---

[46] Baland and Platteau (1996) and Agrawal (2001) provide detailed reviews of this literature in the context of community management of natural resources.

[47] Data presented in chapter 5 indicates that some communities had greater variation in employment than others, which suggests that this proxy of opportunity costs (government employment) may be incomplete. However, the survey failed to capture all forms of community employment in sufficient detail to form more complete proxies.

[48] While data was not available, it is likely that the more indebted the firm the higher the discount rate. For simplicity a constant discount rate is assumed across firms.

ties expected to have higher discount rates[49]. From the household surveys, the proportions of households owning televisions, a mode of transport and holding savings (whether in a bank account or not) before the onset of logging negotiations are also used as indicators for material poverty and hence community discount rates. Community discount rates are also expected to depend on the proportions of households containing a member receiving schooling beyond the compulsory schooling age in Indonesia.

Finally, the community's bargaining power is likely to be higher if the community personally knows its negotiating partner and the more previous experience in negotiations it has. It is also expected to be higher where more households have experience of working for logging companies in the past. The argument here is that these households have greater knowledge of how the logging industry works at the local level and may be well-positioned to power the bargaining process. On the other hand, since these households would have had greater exposure to the environmental impacts of logging, this could also be a proxy for community valuation of the standing forest. In this case, the variable would represent the joint effect of an increase in b and τ, both of which are expected to raise community payoffs.

12.2 Econometric Analysis

Given the small number of observations in the data set (62 communities) and the high degree of collinearity between many of the variables, it was not possible to include a large number of these variables in the analysis. Also, small sample size precluded the use of interaction terms in the following regressions. Appendix 7 reports the two-sided correlation coefficients between the dependent and independent variables for the analyses used to test hypotheses sets 1 and 2, while appendix 8 reports similarly for hypothesis set 3. In most cases the sign of the correlation is as hypothesized.

The analysis for hypothesis set 1 used the probability of obtaining a payment equivalent to or over the threshold Rp. 15,000 per m^3 as the dependent variable, while hypothesis set 2 used the level of financial payoffs (ACTFEE). Hypothesis set 3 used two dependent variables, both ACTFEE and the level of payoffs including social developments (ACTFEEDE). These are illustrated in figure 9.2. Small sample size and high levels of collinearity meant that a formal stepwise regression procedure could not be used. Instead, various combinations of proxies were regressed to obtain the highest possible adjusted R-squared, while still maintaining at least one, preferably two proxies for each model pa-

[49] For example, Pearce and Warford (1993) identify a vicious circle between poverty, high discount rates and environmental degradation.

rameter. This also allowed for the testing of the robustness of the effects of each independent variable on payoffs.

12.2.1 Results for Hypothesis Set 1

In the first stage, a number of proxies are regressed on the likelihood of whether or not the actual payoff falls above or below the threshold of IDR 15,000 per m^3. 19 communities (or 31 percent of the sample) received a fee level that came below this threshold, while the remainder received IDR 15,000 or more per m^3. A sensitivity analysis varying the level of the threshold is conducted to test the robustness of the results to this assumption (see later). Given the discrete nature of the dependent variable and as indicated in chapter 3, a logit model is used. Various combinations of proxies were used, with at least one fitted for each factor in the model at a time. Table 12.1 below gives the results of the proxies that are identified as having the least collinearity effects and which were consistently showing a relatively significant and robust impact on the probability of firm compliance to the contract.

A model is constructed using the following variables. First, for the standing value of the forest to the community (b), AVINCFP (the average proportion of household incomes derived from forest products), and for timber profitability (v), ACTHA (actual area logged by firm) and LOGGEDB4 (the likelihood that the forest has been logged before). In this model, four proxies are included for self-enforcement costs (s): PGOVJOB and DISTMARK (proportion of households employed by government and distance to market, both of which proxy for opportunity costs of time); HPARTORG (household participation in community organizations, collective action ability of community), and; PHHDOMGP (proportion of households belonging to the dominant ethnic group; again, collective action ability of community). Finally, HBANKACC (proportion of households holding savings) is included as a proxy for community discount rate (r^C). Following model specifications in chapter 3, no proxies for community bargaining power (τ) are included in this analysis. The results reported in table 12.1 are quite robust to model variations, in spite of the problems of collinearity and the signs of the coefficients are as expected.

The results in table 12.1 indicate that the probability of a community being able to enforce property rights is increasing in the probability of the community's forest being logged before[50], and decreasing in the size of the area

[50] Note that in preliminary regression models, KUTAIBAR also registered a very strong effect on the probability of whether a community enforces property rights or not suggesting that the location of timber markets may be significant. However, since KUTAIBAR is highly correlated with a large

logged. The latter proxy for timber profitability is significant at the 0.10 level, while the former is significant at the 0.05 level. Following the theoretical model, these results suggest that higher forest quality and larger concession areas increase timber profitability thus increasing the firm's incentives to fight and log unilaterally.

Table 12.1 Econometric Results on Probability of Community being able to enforce Property Rights (Hypothesis Set 1)

Variable (Parameter proxied)	Coeff	Std. Error	t-stat.	P[|T|>t]
Constant	-3.2680	2.6455	-1.235	0.2167
AVINCFP (*b*)	0.1587	0.2272	0.698	0.4850
LOGGEDB4 (*v*)	2.1276	1.1978	1.776	0.0757*
ACTHA (*v*)	-0.009559	0.004437	-2.154	0.0312**
PGOVJOB (*s*)	-0.7598	0.5081	-1.495	0.1348
HPARTORG (*s*)	0.7434	0.3270	2.273	0.0230**
PHHDOMGP (*s*)	0.3034	0.2534	1.198	0.2311
DISTMARK (*s*)	-0.2566	0.1570	-1.635	0.1021
HBANKACC (r^C)	0.1267	0.5515	2.298	0.0216**
No. of observations		62		
Restricted log likelyhood		-38.2		
Chi-squared		37.8		
% of outcomes predicted correctly		71.0		

Note *significant at 0.10 level; **significant at 0.05 level. All results corrected for heteroskedasticity

The probability of a community being able to enforce its rights is increasing with household participation in community organizations and decreasing with distance to market. For these proxies, the former is significant at the 0.05 level, while the latter has a weaker effect but suggests that the further away the market the higher are opportunity costs to the community. The other proxies for self-enforcement costs included in this model, PGOVJOB and PHHDOMGP, have weaker effects than DISTMARK. The proportion of households holding savings before the onset of negotiations has a significant effect (at the 0.05 level) on the ability for communities to enforce its property rights. The proxy for community value of the forest, AVINCFP had a very weak effect in this model. Other proxies for this variable had equally weak effects on the probability of the community obtaining a level of payment at or above IDR 15,000 per m^3. Unlike other

number of variables including AVINCFP and LOGGEDB4 this could not be included in the regression models.

proxies for community value of the forest, AVINCFP is included due to not being highly correlated with other variables in the model and having strong effects relative to the other proxies. A relatively high proportion of outcomes, over 70 percent, are predicted accurately by the model.

The econometric analysis reported here was repeated, for alternative thresholds ranging between IDR 5,000 and 30,000 to test for robustness of the results. The proportion of household incomes derived from forest products (AVINCFP) and the proportion of households belonging to the dominant ethnic group (PHHDOMGP) are influenced by the presence of outliers, which affect the results around the IDR 19-21,000 per m^3 range. The significance levels of all the other variables are reasonably robust.

In conclusion for hypothesis set 1, the probability that the community effectively shares in the benefits from logging is significantly decreasing in the profitability of timber harvesting (v, proxied by LOGGEDB4 and ACTHA), community blockading costs (s, proxied by HPARTORG and DISTMARK) and the community's discount rate (r^C, proxied by HBANKACC).

12.2.2 Results for Hypothesis Set 2

In order to test for the determinants of the actual fees paid to communities, an OLS is constructed, in which only those communities that had received a fee of 15,000 IDR per m^3 or greater are included. Again, at least one proxy is fitted, if not two, to each parameter in the model in order to try and separate out effects. There are no proxies for community enforcement costs (s) in testing for the determinants because the theoretical model suggests that these are only relevant for testing for the probability that a community received a payment equal to or above the threshold. Thus, PGOVJOB, HPARTORG, PHHDOMGP and DISTMARK are all dropped from this model. Proxies for community bargaining power (τ) are incorporated as suggested by the model. Hence, the proportion of households with previous experience of working in the logging industry (EXPLOGCO), and previous community experiences of negotiating with HPH companies (NEGHPH) are included in the model. The results can be seen in table 12.2.

From table 12.2, it can be seen that due to even greater collinearity problems (from the sample being reduced from 62 to 43 communities), more care had to be taken with respect to selecting proxies for the model. The adjusted R-squared for this model shows that only a quarter of the variation in the model is explained by these six variables.

Table 12.2 Econometric Results on Determinants of Actual Fee Levels Paid (Hypothesis Set 2)

Variable (Parameter proxied)	Regressed on ACTFEE Coeff	Std. Error	t-stat.	P[|T|>t]
Constant	10463.0721	10821.134	0.967	0.3400
AVINCFP (b)	307.937	154.535	1.993	0.0539*
EXPLOGCO (b, τ)	214.946	98.911	2.173	0.0364**
LOGGEDB4 (v)	6227.902	6943.740	0.897	0.3757
ACTHA (v)	-0.935	3.680	-0.254	0.8008
HBANKACC (r^C)	170.927	172.212	0.993	0.3276
NEGHPH (τ)	6698.975	6947.731	0.964	0.3414
Dep. Variable	ACTFEE: mean = 41949; SD = 22289			
No. of observations	43			
Deg. Freedom	36			
R-squared	0.35336			
Adj R-squared	0.24558			
Durbin-Watson	2.2812			

Note *significant at 0.10 level; **significant at 0.05 level. All results corrected for heteroskedasticity

From table 12.2, the actual fee received by the community seems to be increasing in the proportion of household incomes derived from forest products (AVINCFP) and the proportion of households with previous experiences with working for logging companies (EXPLOGCO). The first variable is significant at the 0.10 level, while the second variable is significant at the 0.05 level. The quality of the forest (LOGGEDB4) and actual area of forest logged (ACTHA) both have much weaker effects on actual level of fees compared to their effects on the probability of a community able to enforce it's property rights. While the sign of the coefficient for HBANKACC was as expected, the signs for ACTFEE and LOGGEDB4 did not reverse as predicted by the theory, when moving from stage 1 to stage 2 in the model.

In conclusion for hypothesis set 2, for those communities that are able to self-enforce property rights over the forest, payoffs are significantly increasing in the value of the forest considered by the community (b, proxied by AVINCFP) and the community's bargaining power *vis-à-vis* the firm (τ, proxied by EXPLOGCO). The profitability of logging (v) and the community's discount rate (r^C) have much weaker and insignificant effects on community payoffs.

12.2.3 Results for Hypothesis Set 3

In this section, hypothesis set 3 on the total effects of parameters on outcomes is tested. Tables 12.3 and 12.4 below give the results of the proxies that are identified as consistently showing a relatively significant impact on outcomes while explaining up to 45 percent of observed variation in the outcomes. The proxies used to test this hypothesis set follow the ones used to test parameter effects in hypothesis sets 1 and 2. To illustrate the methodology in arriving at these results, table 12.3 shows the OLS model described while table 12.4 gives the same model but also including whether or not communities had previous experiences of negotiating with logging companies in the past (NEGHPH). The results reported here are quite robust to model variations.

The regression results support the hypotheses derived from the theoretical model. The results in table 12.3, for the regressions not including NEGHPH, indicate that the payoffs to the community, using both measures (with or without social developments), are increasing in the proportion of household incomes derived from forest products (AVINCFP), household participation in community organizations (HPARTORG), and the proportion of households holding savings or bank accounts prior to negotiations (HBANKACC). The first two variables are significant at the 0.10 level, while the latter is significant at the 0.05 level. Payoffs are decreasing in the community's distance to the market, which is also significant at the 0.05 level. This supports the hypothesis that the closer the market, the lower the opportunity costs of participating in collective action. The direction of the effect on PGOVJOB, which is significant at the 0.10 level when only financial payoffs are considered, also supports the hypotheses that an increase in the opportunity cost of time leads to reduced payoffs. When using ACTFEEDE as the dependent variable, the significance level drops to around 16 percent, however. The proportion of households belonging to the dominant ethnic group, which is expected to be positively related to the community's ability for collective action in a potential blockade and thus to increase community payoffs, is also significant at the 0.10 level for both payoff measures.

The actual area of forest logged (ACTHA) also has a very significant effect on community payoffs, with the result suggesting that smaller concession areas lead to higher logging fees received per m^3. This should be contrasted with the result for hypothesis set 1 in which smaller concession areas lead to lower profits thus increasing the probability that the community can enforce property rights over the forest. It should be noted, however, that the inability to separate out the effects of timber profitability on payoffs and of concession size on timber profits implies that caution is necessary in drawing any firm conclusions about the reasoning behind this result for ACTHA. The remaining variables in the table 12.3 regressions, the proportion of households with previous experiences of working for logging companies (EXPLOGCO) and forest quality

(LOGGEDB4) seem to have weaker effects. The direction of effect for EXPLOGCO is as expected. The effect of LOGGEDB4 is insignificant, which is not unexpected. Again, from the theoretical model, the effect of *v* on total payoffs is ambiguous because there are two counteracting effects.

Table 12.3 Econometric Results for Total Effects Model (Not Including NEGHPH)

Variable (Parameter proxied)	Regressed on ACTFEE Coeff	Std. Error	t-stat.	P[|T|>t]	Regressed on ACTFEEDE Coeff	Std. error	t-stat.	P[|T|>t]
Constant	-3488.9	12924	-0.270	0.7883	-3046.2	13047	-0.233	0.8163
AVINCFP (*b*)	263.99	153.60	1.719	0.0915 *	270.69	152.13	1.779	0.0809 *
EXPLOGCO (*b*, *τ*)	157.76	112.26	1.405	0.1657	154.29	111.68	1.382	0.1729
ACTHA (*v*)	-2.8604	0.6420	-4.456	0.0000 **	-2.9339	0.6430	-4.563	0.0000 **
LOGGEDB4 (*v*)	7943.1	6022.8	1.319	0.1929	7827.3	5936.3	1.319	0.1930
PGOVJOB (*s*)	-507.99	288.51	-1.761	0.0841 *	-439.30	305.97	-1.436	0.1569
HPARTORG (*s*)	156.59	78.664	1.991	0.0517 *	143.28	77.722	1.844	0.0708 *
PHHDOMGP (*s*)	139.80	76.568	1.826	0.0735 *	151.28	76.619	1.975	0.0535 *
DISTMARK (*s*)	-176.01	51.056	-3.447	0.0011 **	-171.24	49.821	-3.437	0.0012 **
HBANKACC (r^C)	358.98	174.55	2.057	0.0447 **	356.47	172.97	2.061	0.0442 **
No. of observations	62				62			
Adjusted R-squared	0.44348				0.44588			
Breusch-Pagan chi-sq	20.6607				20.7541			

Note *significant at 0.10 level; **significant at 0.05 level. All results corrected for heteroskedasticity

To illustrate this approach and demonstrate the robustness of some of the variables, table 12.4 shows the results for the model with the addition of NEGHPH.

Table 12.4 Econometric Results for Total Effects Model (Including NEGHPH)

Variable (Parameter proxied)	Regressed on ACTFEE				Regressed on ACTFEEDE			
	Coeff	Std. Error	t-stat.	P[|T|>t]	Coeff	Std. error	t-stat.	P[|T|>t]
Constant	-4755.2	12606	-0.377	0.7076	-4210.8	12766	-0.330	0.7429
AVINCFP (*b*)	267.82	153.49	1.745	0.0869 *	274.21	151.96	1.804	0.0770 *
EXPLOGCO (*b*, *τ*)	158.11	107.65	1.469	0.1479	154.61	107.51	1.438	0.1564
ACTHA (*v*)	-2.7774	0.6084	-4.565	0.0000 **	-2.8574	0.6098	-4.694	0.0000 **
LOGGEDB4 (*v*)	7702.8	6231.2	1.236	0.2220	7606.1	6129.7	1.241	0.2202
PGOVJOB (*s*)	-507.29	287.17	-1.767	0.0832 *	-438.66	304.55	-1.440	0.1558
HPARTORG (*s*)	184.70	74.108	2.492	0.0159 **	169.15	73.449	2.303	0.0253 **
PHHDOMGP (*s*)	110.34	73.537	1.500	0.1396	124.17	73.533	1.689	0.0973 *
DISTMARK (*s*)	-168.21	51.688	-3.254	0.0020 **	-164.06	50.474	-3.250	0.0020 **
HBANKACC (r^C)	371.14	178.11	2.084	0.0421 **	367.66	176.55	2.082	0.0422 **
NEGHPH (*τ*)	7156.5	5906.2	1.212	0.2311	6586.5	5863.8	1.123	0.2665
No. of observations		62				62		
Adjusted R-squared		0.44469				0.44530		
Breusch-Pagan chi-sq		20.5382				20.7258		

Note *significant at 0.10 level; **significant at 0.05 level. All results corrected for heteroskedasticity

In table 12.4, it can be seen that the adjusted R-squared does not really change and the results for AVINCFP, EXPLOGCO, ACTHA, LOGGEDB4, PGOVJOB (including the change from using ACTFEE and ACTFEEDE as the independent variable), DISTMARK and HBANKACC are unchanged from table 12.3. HPARTORG becomes more significant from the 0.10 to 0.05 levels and is generally quite a robust variable. Perhaps less robust is PHHDOMGP, which while insignificant when regressed on ACTFEE has a stronger effect (at the 0.10

level) when regressed on ACTFEEDE. This may be due to slight collinearity (0.18403) between PHHDOMGP and NEGHPH. The effect of NEGHPH, while having the expected sign, is, however, not significant.

In conclusion for hypothesis set 3, community payoffs have been shown to be significantly increasing in the community's value of the standing forest (*b*, proxied by AVINCFP), and significantly decreasing in the community's discount rate (r^C, proxied by HBANKACC) and the community's blockading costs (*s*). The latter is proxied by four variables, all of which have significant effects on payoffs. Blockading costs are high where opportunity costs are high (proxied by PGOVJOB and DISTMARK), where communities have lower levels of social capital (proxied by HPARTORG), and are less ethnically homogenous (PHHDOMGP).

12.2.4 A Comparison of Results

The model presented in chapter 3 is tested empirically using relatively similar sets of variables when investigating the total effects of determinants on outcomes and individual stages. The results in table 12.1 (hypothesis set 1) show that LOGGEDB4, ACTHA, HPARTORG, and HBANKACC all have relatively strong and robust effects on the probability of the community enforcing property rights over the forest. When considering total effects (hypothesis set 3), HPARTORG and HBANKACC both have similarly strong and robust effects on payoffs, regardless of the measure used and whether NEGHPH is included in the regression or not. The signs on the coefficients and hence the direction of effect are consistent for these proxies used in testing hypothesis sets 1 and 3. Due to the inability to separate out the counteracting effects of timber profitability on outcomes in the total effects model, the results for ACTHA and LOGGEDB4 should be treated with a degree of caution.

The results of the determinants on actual payoffs (hypothesis set 2) need to be treated with more caution due to the very small size of the sample. Nevertheless, EXPLOGCO and AVINCFP have relatively strong effects on payoffs equal to Rp. 15,000 per m^3 or more. In the testing of total effects of determinants on payoffs, AVINCFP has a similar effect of payoffs, again, regardless of the measure used and whether NEGHPH is included or not.

In conclusion, the results for the total effects analysis appear to be stronger and more robust than those when considering the model in two stages, with a greater number of determinants consistently showing significant effects on outcomes. The inclusion of data for all 62 communities certainly improves the quality of the results for the determinants of the variation observed in actual levels of payoffs. Moreover, hypothesis set 3 allows for the testing of determinants both on actual payments and payments plus some value of the non-monetary benefits provided by the company. The testing of hypotheses 1 and 2 required the use of a threshold level, which was only possible using data collected on promised fees. Knowledge of firm log production schedules would have allowed for a threshold based on promises of non-monetary developments as well. The contracts containing the lowest negotiated or renegotiated fees all contained non-monetary provisions. Thus, a threshold based on the valuation of these would almost certainly have been higher than a threshold calculated on the basis of promised fees alone.

13 Discussion

This chapter begins with a discussion of the observations made in Part II on negotiations and decision-making, and an analysis of household participation in these. A summary and synthesis of the empirical results from Part III follows, including a discussion of equity considerations about the distribution of benefits. While some of the limitations and problems relating to this research such as the data and fieldwork are incorporated into this discussion, additional notes are included at the end of the chapter.

13.1 Negotiations and participation

Almost all communities surveyed in this study have had prolonged and intensive exposure to the logging industry long before the decentralization reforms of the forest sector. Most of the communities surveyed in Kutai Barat, from those situated near the district capital, Melak, to those hundreds of kilometres away in Long Bagun have experienced the logging industry since the late 1960s. First, many communities on the Mahakam river in Kutai Barat involved in non-mechanised logging before many of the first HPH operations in East Kalimantan became established along this river. Moreover, the influence from the regional capital, Samarinda, is very pronounced. This city and the surrounding area contain a huge timber processing industry with the Mahakam river serving as the main transport route for timber. Numerous communities participating in HPHH agreements had direct contacts to this industry. Thus, while few communities in Kutai Barat directly benefited from HPH operations, the proximity of the timber industry has driven the growth of other markets. For a long time, there has been a lot of trade up and down the Mahakam in all manner of goods and services, and a high level of exposure to coastal markets. This is reflected in a relatively cash-rich local economy and numerous local markets in sub-district capitals and in the larger administrative villages in Kutai Barat.

Malinau on the other hand, has been more isolated than the other two districts, and was only more fully opened up to trade and outside influences in recent years. Moreover, communities had perhaps slightly less prolonged exposure to the logging industry in Malinau and Bulungan compared with those in Kutai Barat, at least until the mid- to late-1970s. Experiences of working in the logging industry were evenly spread, although community members from Malinau and Bulungan tended to travel to Malaysia to work as chainsaw operators and surveyors. Thus, the lack of a comparable timber processing industry in Malinau and Bulungan has resulted in the majority of contacts to the logging trade coming via brokers and the smuggling routes into Sabah.

The IPPK system in Malinau and Bulungan was dominated by a small number of brokers, which formed an essential component of powerful, local timber networks, also involving local government officials and, more recently, members of village elites. In the absence of formal law enforcement these networks were necessary for the implementation and enforcement of contracts. While they involved fewer 'formal' brokers, these networks were highly prevalent in Kutai Barat as well. However, the dominance and cartel-like, price-fixing behaviour of the brokers in deal-making resulted in less variation in negotiated fee payments in Malinau and Bulungan compared with Kutai Barat. Moreover, those communities that had less dependence on outside brokers and dealt more directly with contractors may have been able to negotiate for a higher proportion of timber rents than communities that only dealt with outside brokers. In those few cases in Malinau not using brokers, it was telling that these were the ones perceived by other communities in Malinau of leading to the greatest level of net benefits in the district as a whole. Whether considering the 'total fees' received by communities or the so-called 'community fees' earmarked to the wider community, communities in Kutai Barat on average appeared to negotiate higher fees than those in Malinau and Bulungan. Thus on average, by negotiating for and perhaps retaining a greater share of rent within the community, a greater proportion of communities in Kutai Barat may have been 'better-off' overall, although perhaps at the cost of greater inequality between ordinary community members and the village elite.

These inequalities should perhaps be balanced off against the fact that a reduced dependence on outside brokers led to communities, including both the elite and wider community, in Kutai Barat retaining more control over negotiations than those in the other two districts. For example, in Malinau and Bulungan, all IPPK maps and proposed concession areas were supplied to the broker by district forestry offices. Thus, communities had relatively little control over the location and size of the IPPK concession. Furthermore, the semi-nomadic Punan groups in these districts were probably the least aware with respect to concepts of area (as defined by the measurements used in their negotiations with brokers and companies) compared to other groups. By comparison, there were some cases in Kutai Barat in which the company negotiated for a smaller area than indicated by the total size of the permits obtained previously by the community. The implication of permit ownership was that in applying for and owning the permits individually in 100 ha blocks or *petak*, community members in Kutai Barat had a clearer awareness of what they were negotiating away in terms of size, borders and location of concession areas. Thus, there was perhaps a greater sense of control and ownership over the negotiation process in Kutai Barat compared with those in the other districts even while there were similar levels of uncertainty over land claims everywhere. It is possible that this lack of awareness and control may be partly responsible for there being a higher incidence of inter-community conflict over forest claims and borders in communities in Malinau and Bulungan compared with those in Kutai Barat (see next section).

Regarding decision-making over IPPK/HPHH agreements within communities, it was difficult to ascertain how much control and influence most ordinary community members had over the direction of negotiations, particularly given the informal payments that were being made by brokers and companies to community leaders. Moreover, communities had negotiating 'groups', which actually conducted the negotiations with the companies. Thus, the public discussions held in the wider community were not necessarily forums for community decision-making regarding the proposed agreements. Therefore, has participation in these agreements actually led to a 'leveraging' of local democracy as Ribot (2004) suggested it might? Households were asked if there was any public discussion about the IPPK/HPHH that was open to all members, before they were asked whether or not they had actually participated themselves. As described in chapter 7, an average of 86 percent of households per community said there was such a meeting, and a mean 69 percent of households per community actually participated in these meetings. Decisions were made by 'public voting' in around one-fifth of sampled communities only. Instead, decisions were typically made by community leaders, perhaps with some consultation during public meetings. Even where participation and voting occurred, it should be noted that this was almost always undertaken by male household heads and members, with little involvement of women.

Given that the majority of people did not play much of a role in decision-making, what about awareness about the decisions being made on their behalf? From chapter 7, it seems that participation in public meetings did not result in full awareness of the decisions being made or the details of what was being decided. An average of 50 percent of sampled households in each community were aware of the decisions being made (the timing, what steps were to be taken next and so on) and only 38 percent actually had any knowledge of the provisions contained within the agreements being made. Awareness of decision-making was highest in Bulungan (60 percent) and lowest in Malinau (37 percent). Conversely, knowledge of the agreements was highest in Malinau (46 percent) and lowest in Bulungan (30 percent). This suggests that while households in Kutai Barat and Bulungan knew about the decisions that were made with respect to IPPK/HPHH agreements, they did not know what these decisions actually resulted in, i.e. the contents of the agreement. In Malinau on the other hand, households had slightly better knowledge of the agreements made but not how they were made. Thus, active participation did not really translate readily into awareness and knowledge about the process and it seems that the real power still lay with elites in many communities. The responses from Malinau suggest, however, that people may have been slightly less cynical about their influence and participation in the negotiation processes than those in the other two districts. Punan communities that were involved in joint agreements with neighbouring, non-Punan communities, were the ones that were the least involved in negotiations, which was often due to not being treated as 'equal partners' by their neighbours. While active participation may not have led to increased awareness about the deals being made, the relative lack of participation by the Punan (due to being sidelined by others), may have led to these communities having the lowest levels of awareness of all communities in the sample.

13.2 Outcomes and equity

Post-decentralization, all communities received relatively substantial monetary benefits from mechanised logging (see table 13.1). Almost all households received some financial benefits, while a significantly higher proportion of households also received some, non-monetary benefits compared with the pre-decentralization era. Data on the average levels of payments presented in chapter 10 illustrate the huge variation in financial benefits received by households among the three districts. Firms tended to comply much less with promises of social developments than with monetary payments. Furthermore, in many cases the costs of building facilities were taken out of fee payments, sometimes without community members being informed. Communities probably had unrealistically high expectations of what companies were able to provide given that first and foremost the latter were in the business of logging, not the provision of public goods. Nevertheless, IPPK/HPHH companies did make these kinds of promises to communities and there was a kind of understanding among all parties including local government that they would fulfil what would perhaps normally be considered a function of the state.

Relative to the situation before decentralization and despite the inability to value pre-decentralization benefits more precisely, the results suggest that communities benefited significantly in monetary and non-monetary terms from IPPK/HPHH agreements. It should be noted, however, that the actual levels of post-decentralization benefits were still considered by some researchers to be low relative to timber profits (see for example Casson and Obidzinski, 2002, and Limberg, 2004). Profit margins were found to be around 20 percent or more for Malaysian timber investors with capital investments in these areas of East Kalimantan (see Palmer and Obidzinski, 2002), although only four community focus groups in the whole sample had any basic knowledge of timber values.

Table 13.1 Summary of Chapter 11 Results for Impacts from Decentralization

Type of impact		Expected direction of change from decentralization	Results
I. Financial and non-monetary benefits		Increase (+)	Household perceptions of financial benefits received increased significantly (at the 0.05 level) from a mean of 1% of households per community before, to 94% after decentralization. Household perceptions of non-monetary benefits received also significantly increased, from a mean 11% of households per community before to 18% after decentralization.
II. Social	a. Empowerment	Increase (+)	The mean proportion of households indicating community forest ownership alone increased from 21% before decentralization to 82% in 2003-04, a difference which is significantly different from zero at the 0.05 level.
	b. Cohesion and trust	Decline (-)	Anecdotal evidence on problems over the distribution of rents and rent-seeking by village leaders and elites suggests that the onset of IPPK/HPHH concessions led to a decline in trust and possible impacts on the community's capacity for collective action (see below).
III. Enforcement costs	a. Conflict with firm	Increase (-)	25 (42%) and 26 (43%) communities claimed to have undertaken activities against firms before and after decentralization, respectively. Overall, 18% and 34% of households participated in enforcement actions before and after, respectively. Only eight of the 26 post-decentralization cases can be considered 'true enforcement', with high household participation rates in six of these cases. The remaining 18 may be opportunistic or useless blockading due to rent-seeking, or 'true enforcement' undermined by rent-seeking. Another three cases of conflicts with firms were reported, which may also be opportunistic, including one where the firm had complied.
	b. Conflict with other community	Increase (-)	33 communities (55%) reported problems with other communities from post-decentralization concessions. Household majorities in 28 cases (47%) indicated similar problems. Households reported no problems whatsoever in only nine communities (15%).

Table 13.1 Summary of Chapter 11 Results for Impacts from Decentralization (Continued)

Type of impact	Expected direction of change from decentralization	Results
IV. Rent-seeking costs	Increase (-)	21 communities (35%) reported distributional problems. Household majorities in 32 cases (53%) reported similar problems. Blockading against companies with low household participation rates (below 50% of households) were present in at least 18 of these cases. Households reported no problems whatsoever in 13 cases (22%).
V. Environmental costs	No difference	There were no significant differences between perceptions of negative impacts before and after decentralization for water quality, flooding or hunting. Improved access to forests post-decentralization resulted in significantly less (at the 0.05 level) negative perceptions for impacts on forest products and farming compared with the pre-decentralization situation.

Note '+' denotes a positive effect on overall benefits from decentralization; '-' denotes a negative effect on benefits (or an increase in costs).

District governments continued to support and issue IPPK/HPHH concessions for political and economic reasons, even after they were de-legitimised by central government in 2000. These concessions were very popular among local people and community participation led to clear feelings of empowerment among many households surveyed in the field. In this survey, empowerment was quantified in terms of 'forest ownership'. When asked about forest ownership, the proportion of households indicating community forest ownership alone significantly increased from 21 percent before decentralization to 82 percent during the period of survey, i.e. after the IPPK/HPHH concessions had already ceased operations (see table 13.1). Compared to the pre-decentralization data, significantly fewer households indicated that other actors such as companies and the government 'owned' the forest in 2003-04. Given that the Ministry of Forestry still has formal ownership over and land title to all of Indonesia's forests, this is an interesting finding that may have policy implications.

In the total sample of 62 communities, 43 (69 percent) reported problems of company non-compliance with community-company agreements, and 26 of these took self-enforcement action against their respective firms. Comparing the data from before and after decentralization for the sample of 60 communities, the results in table 13.1 seem to indicate that contrary to the expected direction of effect (in chapter 3), there appears to be only a slight increase in the proportion of communities that participated in community-company conflicts, from 42 to 43 percent. Even though many of the actions undertaken before decentralization were not necessarily successful, there were still more enforcement costs incurred by communities than expected.

Post-decentralization, the weaknesses and corruption inherent in local governance meant that communities themselves were responsible for contract enforcement, and some were more successful in undertaking this than others. This in turn led to a variation in actual benefits from community to community. Hence, the empirical evidence shows that while some of these communities have successfully claimed their share of logging rent, others continued to lose their forests to industrial interests for very little financial compensation. Chapter 12 aimed at shedding some light on the reasons underlying the observed variation in community monetary and non-monetary payoffs. The game-theoretic model suggests that in a situation like the one in post-decentralization Indonesia, where community rights remain weak and poorly enforced, the community's ability to self-enforce its rights over the forest is crucial for claiming a significant share of logging profits.

Combining conflict and bargaining theory, the model presented in chapter 3 yields hypotheses on the factors determining the community's ability for self-enforcement as well as other determinants of community payoffs. The theoretical results are supported through empirical analysis using the data described in Part II, especially the more robust results for the consideration of the total effects of determinants on outcomes (hypothesis set 3). From chapter 12, the testing of hypothesis set 1 shows that, as expected, the probability of the community being able to enforce property rights is significantly decreasing in timber profitability. Higher timber profits increase the incentive of the firm to fight and its willingness to log unilaterally. Here, timber profits were higher where concession areas were large and contained forest of a higher commercial quality. The probability of the community being able to enforce property rights is also significantly decreasing with community discount rates (proxied by the proportion of households holding savings) and significantly increasing in social capital (the level of household participation in community organizations). For those communities able to enforce their property rights over the forest (hypothesis set 2), the smaller sample gives a relatively weak set of results. However, as expected, actual payments are found to be significantly increasing in the community valuation of the standing forest (proxied by the proportion of household incomes derived from forest products), and in community bargaining power (proxied by previous experiences of working for logging companies). While timber profitability had an insignificant effect on actual payments, the signs on the coefficients did not reverse as predicted by the theory.

The results for the testing of hypothesis set 3 (total effects of determinants on payoffs) show that communities with a higher valuation of the forest, in particular those that derived a larger proportion of their income from the forest, were more likely to obtain higher payoffs. This is because communities that value the forest more are more willing to fight as their livelihoods are in greater danger from logging damages. The same is true for wealthier communities, which had lower discount rates leading to greater ability to self-enforce rights and also greater bargaining power in negotiations. On the other hand, communities for whom self-enforcement was very costly—because they had high opportunity costs of time or low ability for collective action—were less able to claim an effective share in logging benefits. The results are consistent with collective action theory in that greater ethnic homogeneity and social capital in the form of existing organizations appear to be associated with higher community payoffs. Participation and regular meetings allowed community members opportunities to compare experiences and exchange ideas and information. The impact of changes in the profitability of logging on actual payoffs is more complex when considering the total effects of determinants. While higher profits increase the size of the cake to be shared, they also enhanced the firm's incentives to exploit the community through unilateral logging. Empirically, it was found that communities were able to obtain higher payoffs where concession areas were relatively small.

The results from the community and household surveys presented in table 13.1 indicate that conflicts between communities over land claims and borders occurred in around half of the sample post-decentralization. Though many of these conflicts originated in the past, before decentralization, the freer political environment and sudden rush to capitalise on forest claims through IPPK/HPHH deals caused these to erupt into the open. These conflicts, in which further enforcement costs were incurred, indicate the general weakness and uncertainty of community property rights. Some of the most serious inter-community problems involved the Punan communities, which given their semi-nomadic status, had the most problematic and contestable land claims. Other problems with land claims were observed with the increase in claims since decentralization from powerful, local families such as *pewaris* ('inheritors'). These families complicated the issue of land claims, which were sometimes exacerbated with the influence that they exerted over local government officials. In addition to the *pewaris* claims, many negotiations were complicated further by multiple landholdings and the role of brokers.

The enforcement of property rights through community-firm and inter-community conflicts implies real costs for the community, at least for the substantial proportions of communities that involved in such enforcement activities (see table 13.1). In those communities, perhaps a third of households on average per community participated in self-enforcement activities against companies, double the proportion that participated before decentralization. However, the enforcement of contractual provisions such as environmental rules was patchy at best, and at the very minimum, communities were only able to check logging production but not firm behaviour in the field. This implies high monitoring and enforcement costs, which meant that the first and sometimes the only priority of many communities was to ensure that the companies complied with monetary promises. Therefore, actual compliance with contractual provisions is expected to be a lot lower than was observed in the field because many complaints from village leaders tended to focus first and foremost on monetary benefits.

The sudden influx of cash into materially poor communities led to a surge in rent-seeking activities, usually by members of well-placed village elites, some of whom had been co-opted into local logging networks. Given the nature of the survey, rent seeking and rent capture could not be observed directly in the field, although this survey attempted to qualify, if not directly measure these costs, through questions on the distribution of logging rent within the community. Higher levels of intra-community conflict as a result of problems about the distribution of logging rent coupled with qualitative information gathered from households seems to indicate that rent-seeking was a serious problem in at least half of all surveyed communities, with concomitant unproductive activity costs (see table 13.1). The relatively small size of many of these communities and the relative lack of mobility of many members imply long-term consequences for social capital such as a breakdown of trust and collective action among community members. In communities where these problems were most acute, bitterness, anger and jealousy were apparent from interviews with households and as noted in table 13.1, may have long-term social consequences.

Community and household data were combined to distinguish between 'true' enforcement actions and those that may have been driven by rent-seeking ('opportunistic blockading' and 'useless blockading'). The factor analysis in chapter 11 seems to suggest that 'true' actions are more likely to have occurred in communities with higher household participation and longer duration of action. These actions were expected to have occurred with less frequency, which the results of the factor analysis did not support. However, in a number of cases it was very difficult to distinguish among the actions with the data that was collected. Some actions took place simultaneously, e.g. demonstration and roadblocks. Also, where multiple actions existed, it is clear from some household survey answers, that community members themselves found it difficult to distinguish among the motivations and activities of different groups within a single community. Nevertheless, it seems that opportunistic blockading occurred only in relatively few cases. It might be expected that the relative ease of shutting down company activities, particularly where there was firm non-compliance would have given groups within the community the incentive to 'go it alone'. In which case, larger numbers of small groupings participating in multiple actions would have been observed as 'high' household participation. But the fact that this would have led to a high intensity of intra-community conflict would have distinguished these actions from a single, large group participating in collective actions. The empty quadrant in the top-right of Figure 11.4 shows that this incentive did not appear to exist in the communities surveyed.

From table 13.1, 26 communities (43 percent) reported incidences of company non-compliance with IPPK/HPHH agreements, with some level of household participation in community-firm conflict. Conflicts with companies also took place in another three communities including one where the firm complied with its contractual obligations. These three cases were not indicated as self-enforcement actions in the community surveys and hence, were all almost certainly 'opportunistic', rent-seeking actions. These three cases aside, the confusion given in the responses of many community level and household surveys did not always lend itself to an ability to distinguish between different types of conflict and participation by different groups and hence identify 'true' community collective action. This problem is compounded by the fact that any 'opportunistic' or 'useless' actions might have provided a disincentive for companies to comply further with their agreement. Consequently, this 'muddied' the data and precludes the use of company compliance data for distinguishing among the possible community motivations underlying actions taken against companies where intra-community problems due to rent-seeking were very prevalent. Therefore, with the exception of those eight communities recording low intensities of intra-community conflict, it cannot be concluded whether or not the remaining communities recording household participation in community-company conflicts actually undertook 'true' self-enforcement activities against firms. It seems likely, however, that had they taken place, rent-seeking in whatever form undermined the incentive for a majority of households in these communities to participate anyway. This is supported by the data for fee monies received, as reported in chapter 11. Those communities in which there was firm compliance and zero household participation in any community-company conflicts received Rp. 43,322 (USD 4.81) per m^3, including the value of non-monetary developments, while those that appeared to attempt self-enforcement (whether 'opportunistic blockading', 'useless blockading' or simply undermined by rent-seeking) only received Rp. 24,405 (USD 2.71) per m^3.

Rent capture aside, decentralization clearly led to more logging rent remaining with ordinary community members through more formal rent distribution systems. However, has there been an equitable distribution of these rents? As described in chapter 10, fee monies were typically received by members of the village elites and village leaders in almost all surveyed communities. Furthermore, systems of distribution of fee monies were present in all communities, with systems by 'class' established in almost 40 percent of cases, while the remainder simply divided up the fee money equally either per person or per household. Households were asked about their knowledge of these systems, in addition to the amounts received and whether or not they considered this to be 'fair'. Despite the subjective nature of 'fairness', this question attempted to focus households' minds on how they benefited from logging fees. People often confused this question with the wider issue of the system of rent distribution in the community. This may be due to peoples' sense of community and hence, an inability to separate personal gain from that of other people. The proportions of households in each community answering positively to the question of fairness are plotted against intensities of intra-community conflict in figure 13.1.

From figure 13.1, the regression line seems to indicate that the higher the household perception of 'fairness', the lower the intensity of intra-community conflict over the distribution of fee monies. 'Fairness' is also highly correlated with intensity of intra-community conflict ($p = 0.000$), which seems to confirm that people could not distinguish between solely how their households benefited and the impacts of the fee distribution system on the wider community. Thus, in the wider community people conflated the impacts of fee distribution and sometimes even complained when some of the fee payments were used to cover what might have been legitimate costs accrued by individuals in the community, such as those for permit application. This suggests that people were only content if everyone in the community received the same, regardless of whether they played a productive role in negotiations on behalf of the wider community, or not. In figure 13.1, 34 communities distributed rent on a per household or person basis, while the remainder had a class system. However, the type of system applied does not seem to indicate any clear pattern in relation to 'fairness' or intra-community conflict. On average, 'fairness' was indicated by 57 and 60 percent of households per community where there were 'equal amounts' and 'class' systems, respectively. Intra-community conflict was indicated by an average of 44 percent and 47 percent in communities with 'equal amounts' and 'class' systems, respectively. Thus, neither type of system was perceived as being fairer nor perhaps more transparent than the other.

In Kutai Barat, there was on the whole greater transparency in listing the breakdown of the distribution of promised fees within agreements. However, as noted previously, only a minority of households actually had any knowledge of the contents of agreements and by extension, the distribution of rents. Moreover, even the inclusion of the fee structure in negotiated agreements did not prevent problems of rent-seeking nor did it stop community members from complaining of a lack of fairness with respect to distribution of rents.

Comparing environmental impacts before and after decentralization, there were, as expected, no significant differences between perceptions of negative impacts from logging on water quality, flooding or hunting (see table 13.1). There were, however, significantly less perceived negative impacts on the collection of forest products and farming from IPPK/HPHH operations compared with HPH operations before decentralization. Overall, it appeared that local use ecological damages remained pervasive from IPPK/HPHH concessions, although it is unclear from a survey of this type whether the impacts from these compared with those from before decentralization have intensified or not. The measure of damages according to peoples' perceptions is obviously limited in that it does not say anything about relative environmental impacts. Moreover, budget and time constraints also meant that these perceptions could not be cross-checked with actual measurements of damages from the field. In areas, where several communities may be located and where concessions are many and overlapping, discussions with community members suggest that logging impacts have gotten worse over time. This is particularly the case in almost half of the sample where the IPPK/HPHH concessions were operating in previously logged areas.

Figure 13.1 'Fairness' and Intra-community Conflicts from Systems of Fee Distribution in Communities.

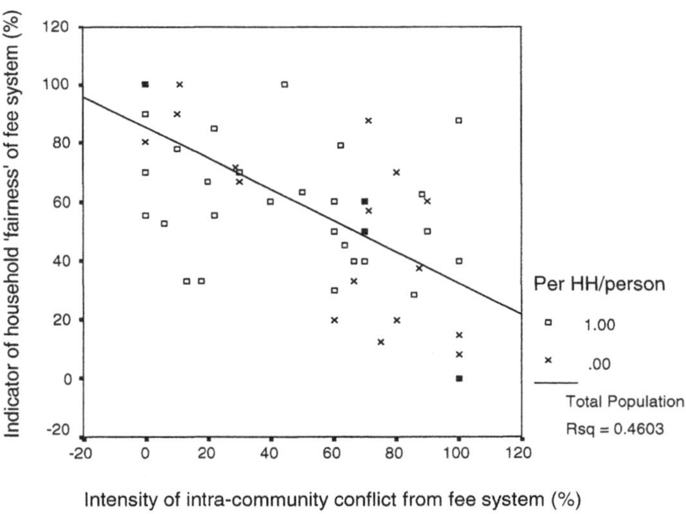

Note 1 = equal amount per household or per person; 0 = class system.

The ability of many communities to have some say in where concessions were to be located, in addition to insistence on logging rules may have tempered the direct, immediate impacts with respect to the communities themselves. There is, however, little statistical evidence of a trade-off between environmental and financial provisions. It seems that communities could negotiate relatively high fees at the same time as environmental provisions. This finding provides further support to the theoretical model in chapter 3, that the ability of communities to establish *de facto* property rights over the forest is a prerequisite to obtaining any kind of benefits at all, including the minimizing of environmental damages from logging activities. Finally, it should be stressed that in addition to being unable to measure actual damages, perceptions of the overall damages from these concessions are also unknown; community non-use values and global damage costs from logging, both before and after decentralization have not been considered due to being outside the scope of the survey.

Overall, while it is not possible to value many of these costs and benefits in order to construct a definitive cost-benefit analysis, the conceptual framework developed for the purpose of assessing the impacts from decentralization can be used to identify and analyse the variation in impacts. It has been shown that there is a lot more variation in types of impacts after decentralization compared with the situation before decentralization. There are communities that have clearly benefited well from the small concessions regime. For example, all three communities in Memahak Besar negotiated agreements that generated a lot of money (an actual fee of at least Rp. 60,000 or USD 6.67 per m^3 was received in each community including value of developments), some of which was used to help develop an electricity network and a new drainage/sewerage system, in addition to an agricultural plantation developed by the firm. Despite acknowledging the high ecological cost of situating concession areas within an old HPH area that had already been logged three times since the late 1960s, the feeling of empowerment among community members was tangible.

Another positive example is Sekatak Bengara, which reported no problems whatsoever with its company, i.e. there was full firm compliance with its contractual obligations, nor with its neighbouring communities, nor from within the community. Although the fee of Rp. 30,000 (USD 3.33) per m^3 that was received was not particularly high, all monies were distributed directly to community members, which prevented the possibility of the village elite seeking rents from payments that were destined for the rest of the community. Moreover there was apparently some plantation development, although this may have had nothing to do with the logging company and may have been developed separately by the local government. Also, unlike Mamahak Besar, local ecological impacts were minimised due to the siting of the concession area downstream of the community and away from hunting and forest product areas. This, however, was the case with the pre-decentralization HPH concession as well, which implies that this may have been less of a negotiated aspect of the agreement and more a matter of good fortune for the community in being located in an advantageous spot on the river, far from any logging activities.

The majority of sampled communities reported numerous problems from post-decentralization agreements. In Bitt Unyang in the administrative village of

Ujoh Bilang, as described in chapter 10, community leaders and well-positioned members did not honour their internal fee distribution agreement made with other members of the community. Instead, they not only kept most of the fee money when it was received, but also moved out of the village, relocated to the provincial capital and built houses with this money. In this community, over 85 percent of households responded that there were conflicts as a result of this behaviour. While there is little evidence of these kinds of intra-community conflicts becoming violent, in this case there was clear evidence of deep divisions and bitterness among community members. Again, as noted in table 13.1, this may have long-lasting impacts on cohesiveness, trust and the ability for collective action within this particular community. Moreover, not only had the leaders failed to distribute fee monies but they subsequently blamed the company for non-compliance despite the fact that the company had already made some payments to the community leaders. Early on, many people in this community suspected that their leaders were capturing and not sharing rents, and although a demonstration did take place against the company, there was a very low (14 percent) rate of participation within the community. The company failed to comply with further payments to the community because by this stage the community was indifferent to taking action against the company. Coupled with rent-capture, an inability to enforce the contract resulted in an actual fee of just under Rp. 9,000 (USD 1.00) per m^3. This is an illustrative example of the third of all sampled communities located in the lower-right quadrant in figure 11.4, where there was firm non-compliance, low community participation to ensure compliance and high rates of intra-community conflict.

While households and village leaders were keen to stress monetary and non-monetary benefits and were, in numerous cases, clearly empowered by the process, they tended to be unaware of the associated costs and impacts: on their relations with one another, other communities and their environment. Communities that benefited well financially and also managed to maintain good relations, both within and with other communities and groups, were rare. For example, Tanjung Lima, Tajan and Selidung collectively bargained a fee level, successfully enforced their contract and received the highest per household payments in the whole in Malinau. There were signs, however, that the communities had started drifting apart, with those who had gained most moving to a new location altogether. There were numerous accusations of corruption towards the fee management team and the local Catholic Church, which were entrusted with managing the fee payments to the communities. Moreover, households had also become less productive and were simply living off their fee payments. This is perhaps understandable given that each household had received the equivalent of thousands of dollars in fee payments in a short period of time, in an area where earnings from the sale of agricultural surpluses averaged less than USD 500 per year. However, from interviews in these communities, many households were simply unaware of where the fee money had originated from and some did not even care, despite all the associated problems. Similar attitudes were encountered in other communities, too.

13.3 Notes on limitations and possible extensions

The surveys used in this study sought to combine quantitative and qualitative techniques. The structure and composition of the questionnaires were designed to be both practical and rigorous, on the basis of the case study approach and a more formal, economic approach. Taken together, the use of community level and household level questionnaires allowed for the use of the community as the principle unit of measurement while information collected from households was used to corroborate and substantiate the community level answers. The use of many, short close-ended questions ("yes", "no", "don't know") allowed for more direct comparisons across communities, although there was a danger that vital information might bypass the interviewers. To get around this problem, the research team was encouraged to talk to as many people as possible in the relatively short time available in each community, and collect information that did not always fit comfortably into the survey format used. Thus, while pure sociological and anthropological research approaches typically involve spending more time in research areas and tend to use more open-ended questions, the methodology used in this study attempted to blend these approaches with a more rigorous and systematic methodology. The use of an economic framework to analyse many of the responses from the surveys required that the research be undertaken in this way.

The objective of the survey was for a greater breadth of data and information, at the cost of some lack of depth. For example, information on most inter-community conflicts was collected on the basis of simple indicators and qualitative details, with little actual quantitative data collected. It is perhaps inevitable that details will be lost in undertaking an approach of this kind, even accounting for the fact that it was always going to be impossible to observe everything that had occurred in the surveyed communities. Thus, the characteristics of communities were captured by quite approximate indicators, e.g. the use of government employment to proxy for opportunity costs. A more detailed survey would aim to capture the variation in employment among communities and hence, allow for the formulation of more precise proxies. In addition to the financial and logistical constraints in undertaking a more detailed survey, it was clear from the fieldwork that more detailed questions often resulted in misunderstandings among respondents, and hence, failed to elicit the clear responses necessary to form more robust proxies. For example, the surveys failed to capture good quality data for household participation and duration of all recorded community-firm conflicts in addition to the longest recorded event. This meant that community enforcement costs could not be valued properly. Hence, there may even be limits to the type and quality of data that can be collected from the field even where these are available.

In making a comparison of IPPK/HPHH experiences with those of mechanized logging before decentralization, the survey asked respondents questions on events that had happened many years previously. In the absence of time-series data, these kinds of questions always run the risk of obtaining answers that have been contaminated with knowledge gained from having the benefit of hindsight. To counter this, the pre-decentralization section was kept as brief as possible. Moreover, like most of the rest of the survey, the questions dwelled less on attitudes, opinions and preferences and instead attempted to draw these out indirectly through data and information gathered on actual events and actions, with corroboration across the two surveys. Where the surveys attempted to elicit attitudes through the presentation of for example, 'alternative scenarios', they inevitably failed. Some of these scenarios had actually occurred in some communities and these questions were eventually abandoned.

Where possible, the information gathered was cross-checked with other sources. In particular, details on the IPPK/HPHH logging agreements were checked with the information that was available from district government, data collected from previous studies and extended informal discussions with the few community members who had worked for the IPPK/HPHH companies. However, it should be noted that for the most part, actual contracts were unavailable to the research team either because they did not have copies in the first place or because they no longer existed. Thus, contractual details were typically obtained from discussions with the community leaders in the focus group discussions, i.e. from what could be remembered and peoples' perceptions, and not always from the actual contracts. The contracts seen by the research team were by and large relatively short and simple, although a few were long, detailed and sophisticated. In the context of little or no government enforcement of formal contracts, the significance of written agreements and even those that had been notarised is likely to be very limited. Moreover, apart from having a symbolic value in being an acknowledgement of rights to the forest, written agreements were of little value in places where the majority of the population have the bare minimum of literacy and numerical skills. Furthermore, since self-enforcement and a dependence on trust and relationships within the local networks were more important than the minutiae of contracts, what was perceived and remembered in relation to the contract was perhaps more important than the contract itself. This is because these were the aspects of the agreement that people within the communities were most prepared to fight for, in establishing *de facto* property rights over the forest.

The considerable financial and logistic constraints on collecting quantitative data from a cross-section of communities limited the sample size and thereby the possibilities for a more detailed econometric analysis. While the empirical results reported in chapter 12 appear relatively robust, high levels of collinearity among variables prevent separating out the effects of alternative factors influencing the broad parameters identified by the model. The model as applied here was restricted to the analysis of financial outcomes alone. There are other outcomes that have been drawn from the data and are briefly summarised in table 13.1 that could be analysed using the same model. For example, the significantly greater perceptions of community forest ownership post-decentralization might relate to experiences of actually standing up for these perceived rights in the past. Strong perceptions of community ownership may have been formed and stronger *de facto* rights may have been claimed long before the onset of IPPK/HPHH negotiations, i.e. they may have been formed as a result of community participation in activities against HPH companies before decentralization. Thus, the model adapted in chapter 3 could be used to test this and other factors underlying the variation in community perceptions of forest ownership.

There are, however, a number of limitations in applying a model such as the ones developed by Engel, López, and Palmer (2006) and Engel and López (2004). First, these do not explicitly include the process of negotiations leading to agreements between communities and companies. A model extension would add another stage to explicitly distinguish between payments negotiated and compliance with agreements, although the fluidity of negotiations and agreements meant that in some cases there was a degree of uncertainty about what actually constituted a 'final' negotiated settlement. Another extension of the model would be to move from the perfect information case to imperfect information. The fact that there were actual incidences of conflict in many of the cases surveyed in East Kalimantan implies that communities and firms probably did not have perfect information with respect to the other's parameters. The data collected on community-company conflicts could be used to test an extended version of the model incorporating imperfect information.

14. Conclusions and Outlook

For the first time in 35 years, forest-dependent communities in Indonesia were given some stake in forestland, and while property rights are still uncertain, communities have been given legal acknowledgement of their claims short of formal land title. The opportunity to trade in customary rights to the forest for a share of timber rents and the subsequent proliferation in IPPK/HPHH contracts was the first step in utilizing this stake. This thesis provided a unique opportunity to assess the impacts of these decentralization reforms on communities through the development of surveys that sought to combine qualitative and quantitative techniques. The impacts of decentralization on communities were hitherto part of a phenomenon that has been mainly researched through the case study approach and more ad hoc research techniques. While acknowledging financial gains and environmental costs to communities from post-decentralization concessions in Indonesia, studies such as Casson and Obidzinski (2002) and Resosudarmo (2004), for example, did not attempt to quantitatively assess these impacts on communities. More generally, research undertaken on the impacts of decentralization in natural resource management has not, until now, really attempted to quantify these effects and compare them to the effects of alternative management regimes and systems of property rights. Instead, case studies are typically compared in order to abstract the main trends and patterns of cause and effect (see for example, Larson and Ribot, 2004).

In this study, a conceptual framework was formulated and the expected directions of effect with decentralization derived for five defined categories of impact. The results from a sample of 60 communities show that communities did indeed benefit financially from decentralization, although with large variation across the sample and among households. In addition, this study analysed the costs of enforcement (*vis-à-vis* companies and other communities) and attempted to assess the many and varied impacts of rent-seeking within communities. Neither of these impacts had been quantitatively researched before, at least not in the Indonesian context. The results show that many communities incurred real costs, which should be set off against the financial and non-monetary benefits. Also, while there were environmental impacts from post-decentralization concessions, these may not be significantly worse than under the centralized regime. Furthermore, this thesis undertook an analysis of the possible trade-offs between the environmental and financial contractual provisions that were negotiated by communities. Again, this is the first time such an analysis has been attempted and the results suggest that there is in fact little evidence for communities trading off one set of provisions against the other. Instead, communities that negotiated environmental provisions in their agreements on average also received higher payments than those that did not. Hence, communities that negotiated contracts containing environmental provisions had a greater capacity to self-enforce agreements than those with no such provisions in their contracts, which implies that community bargaining power and self-enforcement capacity are linked as shown theoretically in the models adopted from Engel, López, and Palmer (2006) and Engel and López (2004). Also, decentralization clearly led to more feelings of empowerment among community members with a significantly higher proportion of households perceiving community ownership over the forest after decentralization compared to the period before. However, there was still relatively little direct household participation in the formation of community-company agreements. It was shown that among households, there appeared to be little awareness and knowledge about these agreements and that village elites were still driving most community decision-making.

The models by Engel, López, and Palmer (2006) and Engel and López (2004) led to the formulation of three sets of hypotheses on the determinants of payments received by communities from post-decentralization logging agreements. The relatively large sample of communities allowed for the first analysis of community-firm interactions to be based on quantitative data, and for the testing of hypothesised effects. In particular, this approach enabled the testing of some of the tenets and findings of collective action theory (see Ostrom, 1990; Baland and Platteau, 1996). Similar to the assessment of decentralization impacts, research relating to collective action theory is typically based on the case study approach thus making direct comparisons problematic. Some key results from the econometric analysis in this study are consistent with collective action theory, such as the influence of social capital and ethnic homogeneity on outcomes. These results are supplemented by a relatively large amount of qualitative information, which provides a human dimension to the quantitative results and has reinforced many of the conclusions from the descriptive statistics and econometric analysis.

As noted already, community-company agreements provided local people with direct monetary benefits and some non-monetary developments in the absence of any (state) alternatives; from scholarships through to the levelling of ground and repairing of churches. Thus, in some cases, these arrangements were very responsive to the needs of local communities, as well as providing retribution and justice for the perceived 'theft' of land by central government prior to decentralization. The fact that many promises made by brokers and companies were broken without there being even more anger and conflict was perhaps more symptomatic of low expectations on the part of many communities than anything else. Furthermore, these agreements were also the first attempt to draw so-called 'extra-legal' logging arrangements into a common, legal framework, perhaps more by default than by design. This allowed for (some) local level taxation and hence, for some rent to be retained by local government.

An IPPK/HPHH contract was derived from a relationship of trust with people embedded within local logging networks. Unenforceable by the state, this kind of agreement was more important than the state 'norm', that is the notion of notarised contracts and permits. While the latter were often present, the existence of weak and uncertain property rights in East Kalimantan meant that they were essentially meaningless without the former. These kinds of contractual arrangements and transactions are the norm for the majority of the world's people living and working in developing countries (De Soto, 2000). In other words, the districts and communities surveyed in East Kalimantan are under-capitalised societies characterised by localised, pluralistic, weak or non-existent formal and standardised property rights and codes of enforceable 'law', located outside what would be recognised as the 'state'. Thus, all the actors involved in IPPK/HPHH agreements were still largely dependent on the institutions that existed before the political and economic upheavals of 1997-98 onwards. Property rights were defined and derived from one or a combination of Suharto's centralised government, the Dutch colonial institutions, the sultanates (e.g. of Bulungan), and customary law. Moreover, the latter institutions are very localised with little recognition from one area to the next. For example, the forest claims of the *pewaris* families were derived from the rights to the harvesting of birds nests that were handed out a long time ago by the Dutch colonial authorities, while the majority of communities claimed forest on the basis of customary law. To complicate matters further, HPH companies still derived authority from central government with concessions issued from the Suharto era. Perhaps the biggest legacy of Suharto's centralised control was to leave a vacuum, into which a confusion of institutions defining various rights and claims have been jostling for control and influence.

Given how *de facto* decentralization occurred following the economic crisis and during Suharto's last days in power, and some time before the decentralization laws were implemented, is it possible that these changes would have happened even if Suharto had stayed in power? Had the forest continued to degrade and had large-scale logging remained uneconomic in many parts of Indonesia, small-scale concessions with or without community involvement may have become the norm, replacing large-scale timber concessions as the main supply for the domestic plywood industry or even international demand. This may have happened eventually, because although Suharto was the figurehead for repressive regime, he may have had to compromise in order to preserve power at some stage. As noted in chapter 11, a significant proportion of communities were scared of his regime. After decentralization, Casson and Obidzinski (2002) noted that the local military were left out of IPPK deals because the concessions did not require any 'protection'. Despite this, there is a lot of evidence for well-established, local timber networks with the involvement of the military, both pre- and post-decentralization (see for example, McCarthy, 2000).

In 2000, the central government decided to de-legitimise IPPK/HPHH agreements because they apparently encouraged 'illegal logging' and discouraged 'sustainable' forest management (see for example, Ministry of Forestry, 2004). However, what constitutes 'illegal' in Indonesia is not very clear and is open to interpretation. For example, before decentralization, the activities of many HPH concessionaires were also technically 'illegal' in that they ignored rules set by the central government, in addition to marginalizing forest-dependent communities (Schwarz, 1990; Casson and Obidzinski, 2002). Since district governments reluctantly phased out the IPPK/HPHH systems, they have been adapting a new concessions system known as the 'HPH mini'. This very approximately combines the HPH system with the IPPK/HPHH systems. The concessions are generally larger than IPPK/HPHH ones and require that communities combine forest claims in order to qualify. The objective is to impose some order on the proliferation of forest claims and hence, inter-community conflict over forest borders. However, the new system is likely to lead to more, not less inter-community conflict in the future. Local governments, anxious to establish the new system as quickly as possible and ensure future revenue streams from log production, are imposing new boundaries on communities and forcing them to group together without the settlement of existing claims. Instead, the acknowledgement of community property rights from IPPK/HPHH agreements should be the first step in the provision of more clearly defined and secure rights, in order to reduce uncertainty and conflicts over claims to forest land. Also, community forest rights need to be extended beyond mere 'use rights' in order to reduce the pro-logging bias.

Communities participating in the 'HPH mini' are also not supposed to negotiate with companies and instead, receive a government-mandated, fixed level of benefits. However, as this system was being gradually introduced in the areas surveyed, it was clear that many people preferred the IPPK/HPHH due to having the opportunity to negotiate relatively freely (see for example, Limberg, 2004). From the local perspective, the new system may have little more credibility than the HPH system, due to being less responsive to the needs of communities. Nevertheless, given the continuing dislocation of local governments from central government and the power of local logging networks, it seems that the old system will simply continue under a different guise but with less rents being negotiated directly by the communities and more remaining with local government. Higher expectations as a consequence of IPPK/HPHH experiences means that many communities may not accept this situation and will probably lobby, at the very minimum, for a greater share in rents.

Relationships between local governments and communities were not particularly strong. On the whole, many communities did not really trust local governments and perceived local officials as being not only incompetent and unresponsive, but corrupt as well. While local governments have been attempting to build new relationships with local people, misunderstandings persist. For example during the fieldwork undertaken in 2003, new systems of local taxation were being instituted in Malinau, which were resented by local communities. Given the relative lack of service provision in the past, this attitude is perhaps understandable. However, the same communities did not seem to realise that their demands for more social developments from the local government depends on its ability to raise tax revenues in the first instance. What is required is a broader effort to deal with these kinds of misunderstandings with the formation of a new social contract between forest-dependent communities and the state.

There is also a need to provide alternatives to any system of logging in places where there is a high demand for social developments, and a high abundance of natural resource rents. Systems of direct Payments for Environmental Services (PES) whereby local communities are paid to act as 'guardians' to the forest may be a feasible alternative to logging. These are market-based mechanisms for resource conservation in which NGOs, governments, or international donor agencies make periodic payments to individuals or groups that supply environmental services (see for example, Pagiola and Platais, 2002). PES can be seen as an alternative type of contract, where user groups or individuals are paid for conserving the resource rather than trading it for rents derived from commercial, short-term exploitation. Thus, PES may ensure that externalities are at least partly internalized by local people in their resource management decisions. However, knowledge of local property rights needs to be supplemented with knowledge on local systems of incentives, in order to deal with powerful, local logging networks and hence, ensure compliance with PES contracts. Given the weakness of the state, NGOs could act the role as the neutral third parties necessary for the successful implementation and monitoring of PES contracts. In Kutai Barat for example, NGOs have had a close relationship with the district government where they have played an important role as partners to the district government and other stakeholders in planning forestry developments (KKPKD, 2001).

The results from this study provide information with regards to how communities value the forest. It also sheds some light on the conditions under which PES schemes might work and at the same time improve community welfare via strengthening their ability to self-enforce property rights over the forest. As expected, poorer communities with higher discount rates received lower payments. Thus, poverty reduction and a lowering of community discount rates through improved access to credit might be conducive to improving the incentive to self-enforce contracts. Moreover, a lowering of the costs to self-enforcement through reducing the opportunity costs of time and enhancing community ability for collective action would improve communities' incentives and payoffs as well. This could be undertaken via the improvement of access to markets and trade, and via training workshops to improve community participation and leadership. The latter activity could also seek to create village organizations where there are none and improve existing ones, hence increasing community social capital. The results show that initiatives, for example to improve market access, might lower opportunity costs. This is contrary to the conventional view that improved market access raises opportunity costs, thus reducing self-enforcement ability. General education and awareness training might also help to improve community awareness of the links between environmental damages and community welfare, thus attempting to raise local values of the forest. Overall, the models developed by Engel, López, and Palmer (2006) and Engel and López (2004) emphasise the importance of the ability to self-enforce property rights in situations where rights are weakly defined and poorly enforced by the state. However, the model also suggests that self-enforcement is a second-best solution in the protection of property rights. Given the costs of self-enforcement, a stronger and more impartial regime of government contract enforcement may be more beneficial for communities in the long-run.

There are a number of factors from the results that do not appear to be amenable to policy advice. For example, it was clear from the fieldwork that many communities had been gradually shifting away from forest dependence to other sources of income, partly due to land change patterns, patterns of migration towards towns and shifting preferences towards the creation of agricultural surpluses, e.g. through the development of agri-business and commercial plantations. The results indicate that this shift could be detrimental in environmental terms as lower community valuations of the forest reduce communities' willingness and ability to fight for *de facto* property rights. Nevertheless, the livelihoods of the overwhelming majority of people in this sample were still very much dependent on land that was being directly or indirectly affected by logging activities. Therefore, raising community awareness to help realize the local value of forests may still be important despite a gradual shift away from direct forest dependence. The improvement of relations between different ethnic groups within communities may also increase the community ability for collective action, especially given constant, dynamic changes in the ethnic make-up of many communities.

Since the late 1960s, what used to be a very remote, heavily forested and relatively inaccessible part of Indonesia has been rapidly opened up to the outside world. Increased mobility and the spread of the media and communication technologies have led to increased awareness of this world. Furthermore, communities in East Kalimantan have become part of a cash economy that was formerly restricted to the coastal areas, and hence, new consumers. People are still very much dependent on the local resource base to sustain this consumption. The demand for and benefits from IPPK/HPHH concessions fits neatly into this picture. This is despite the inability of many communities to self-enforce property rights, thus leaving them as poor as they were before negotiations. As the forests become more depleted and with predictions from the Ministry of Forestry that East Kalimantan will have no commercial forest stocks left by 2010, it is clear that logging in its current form is completely unsustainable. The failure to internalize the environmental damages from commercial, mechanised logging in the past implies long-term, though perhaps unpredictable long-term effects. Furthermore, the social problems and conflicts resulting from the anarchy of the IPPK/HPHH system may also change the whole concept of a 'traditional indigenous community', especially if forest degradation forces people to move away from their ancestral lands.

While it was clear that making these contracts empowered communities after constant marginalization and even repression from Suharto's forestry policies, it is a lot more difficult to get a measure of what this means and what this implies for the future. Before decentralization, only a quarter of nearly 700 sampled households claimed that the community had any kind of ownership over the forest, a figure that rose to an extraordinary 90 percent of households in 2003-04. Furthermore, many community leaders had a tangible confidence about 'their' property rights. These perceptions have implications for future state forest policy, particularly with regards to land reform and the control over and distribution of logging rents. Decentralization could be the first step in the right direction in the sense that community empowerment needs time to come to fruition, although it is questionable whether there is now enough time for this to any effect on local forest management given current rates of forest degradation.

List of References

Agrawal, A. 2001a. The Decentralizing state: nature and origins of changing environmental policies in Africa and Latin America 1980-2000. Conference paper. San Francisco, US: the 97th Annual Meeting of the American Political Science Association, 30 August-2 Sept. 2001.

Agrawal, A. 2001b. 'Common property institutions and sustainable governance of resources', *World Development* 29(10):1649-1672.

Agrawal, A., and J. Ribot. 1999. Accountability in decentralization: a framework for South Asian and African cases. *Journal of Developing Areas* 33 (Summer): 473-502.

Alston, L.J., Libecap, G.D., and B. Mueller. 1999. A model of rural conflict: violence and land reform policy in Brazil. *Environment and Development Economics* 4(2): 135-60.

Anau, N., Iwan, R., van Heist, M., Limberg, G., Sudana, M., and E. Wollenberg. 2002. Negotiating more than boundaries: conflict, power and agreement building in the demarcation of village borders in Malinau. Bogor, Indonesia: Center for International Forestry Research.

Angelsen, A. 2001. Playing games in the forest: state-local conflicts of land appropriation. *Land Economics* 77(2):285-300.

BPS (Badan Pusat Statistik). 2002. *Kalimantan Timur Dalam Angka*. Samarinda, East Kalimantan, Indonesia: East Kalimantan provincial government.

BPS (Badan Pusat Statistik). 2000. *Statistik Perusahaan Hak Pengusahaan Hutan, 2000*. Jakarta, Indonesia: Republic of Indonesia.

Bagian Perekonomian. 2000. *Statistik Pemerintah Daerah Kabupatan Bulungan*, Kalimatan Timur, Indonesia: East Kalimantan provincial government.

Baland, J.M., and J.P. Platteau. 1997a. Wealth inequality and efficiency in the commons Part I: the unregulated case. *Oxford Economic Papers* 49(3): 451-82.

Baland, J.M., and J.P. Platteau. 1997b. Coordination problems in local-level resource management. *Journal of Development Economics* 53: 197-210.

Baland, J.M., and J.P. Platteau. 1996. *Halting degradation of natural resources. Is there a role for rural communities?* Rome, Italy: Food and Agriculture Organization of the United Nations (FAO), and Oxford, UK: Clarendon Press.

Bappeda Tk II Bulungan. 1998. *Ikhtisar Statistik Kabupaten Bulungan 1997*. Samarinda, East Kalimantan, Indonesia: Kantor Statistik Kabupaten Bulungan, CV Krama Agung.

Bardhan, P. 1993a. Analytics of the institutions of informal cooperation in rural development. *World Development* 21(4): 633-39.

Bardhan, P. 1993b. Symposium on management of local commons. *Journal of Economic Perspectives* 7(4): 87-92.

Barr, C., and K. Obidzinski. 2003. The effects of decentralisation on forests and forest industries in Berau District, East Kalimantan. Bogor, Indonesia: Center for International Forestry Research.

Barr, C., and I.A.P. Resosudarmo. 2002. Decentralization of forest administration in Indonesia: community livelihoods and economic development. Bogor, Indonesia: Center for International Forestry Research.

Barr, C., Wollenberg, E., Limberg, G., Anau, N., Iwan, R., Sudana, M., Moeliono, M., and T. Djogo. 2001. The impacts of decentralisation on forests and forest-dependent communities in Malinau district, East Kalimantan. Decentralisation and Forests in Indonesia Series: Case Study 3. Bogor, Indonesia: Center for International Forestry Research.

Bose, P. 1998. Formal-informal sector interaction in rural credit markets. *Journal of Development Economics* 56 (2): 265-80.

Bray, D., Merino, L., Negreros, P., Segura, G., Torres, J.M., and H. Vester. 2003. Mexico's community-managed forests as a global model for sustainable landscapes. *Conservation Biology* 17 (3): 672-677.

Brown, D. W. 1999. Addicted to rent: corporate and spatial distribution of forest resources in Indonesia; implications for forest sustainability and government policy. Report No. PFM/EC/99/06. Jakarta, Indonesia: Indonesia-UK Tropical Forest Management Programme (ITFMP).

Burton, P. S. 2004. Hugging trees: claiming *de facto* property rights by blockading resource use. *Environmental and Resource Economics* 27 (2): 135-163.

Carney, D. 1995. Management and supply in agriculture and natural resources: is decentralization the answer? *ODI Natural Resource Perspectives* 4 (June). London, UK: Overseas Development Istitute (ODI).

Carney, D., and J. Farrington. 1998. *Natural resource management and institutional change*. London: Routledge.

Casson, A. 2001. Decentralisation of policies affecting forests and estate crops in Kutai Barat district, East Kalimantan. Case Studies on Decentralisation and Forests in Indonesia, case study No. 4. Bogor, Indonesia: Centre for International Forestry Research.

Casson, A., and K. Obidzinski. 2002. From New Order to regional autonomy: shifting dynamics of 'illegal' logging in Kalimantan, Indonesia. *World Development* 30(12): 2133–2151.

Chambers, R. 1997. *Whose reality counts? Putting the first last*. London, UK: ITDG Publishing.

DKDA (Dinas Kehutanan Dalam Angka). 1999/2000. *Forestry Department statistics*. Samarinda, East Kalimantan, Indonesia: Department of Forestry.

DKKT (Dinas Kehutanan Kalimantan Timur). 2000. *Forestry Department statistical handbook*. Samarinda, East Kalimantan, Indonesia: Department of Forestry.

DKKT (Dinas Kehutanan Kalimantan Timur). 2002. *Forestry Department statistical handbook.* Samarinda, East Kalimantan, Indonesia: Department of Forestry.

Dermawan, A., and I.A.P. Resosudarmo. 2001. Forests and regional autonomy: the challenge of sharing the profits and pains. In *Which way forward? People, forests, and policymaking in Indonesia.* Colfer, C.J.P. and P. Resosudarmo (eds.): Chapter 15. Washington DC, US: Resources for the Future, and Bogor, Indonesia: Center for International Forestry Research and the Institute for Southeast Asian Studies (ISEAS).

De Soto, H. 2000. *The mystery of capital: why capitalism triumphs in the West and fails everywhere else.* New York, US: Basic Books, Pereus Books Group.

Dixit, A., and B. Nalebuff. 1991. *Thinking strategically.* New York, US: Norton.

Dudley, R.G. 2001. Dynamic of illegal logging in Indonesia. In *Which way forward? People, forests, and policymaking in Indonesia.* Colfer, C.J.P. and P. Resosudarmo (eds.): Chapter 16. Washington DC, US: Resources for the Future, and Bogor, Indonesia: Center for International Forestry Research and the Institute for Southeast Asian Studies (ISEAS).

Edmunds, D., and Wollenberg, E. 2003. *Local forest management: the impacts of devolution policies.* London, UK: Earthscan.

Edmonds, D., Wollenberg, E., Contreras, A., Dachang, L., Kelkar, G., Nathan, D., Sarin, M., and N. Singh. 2003. In *Local forest management: the impacts of devolution policies.* Edmunds, D., and E. Wollenberg (eds.): 1-19. London, UK: Earthscan.

Engel, C., and C. Palmer. In press. Who owns the right? The determinants of community benefits from logging in Indonesia. In *Forest Policy and Economics.*

Engel, S., López, R., and C. Palmer. 2006. Community-industry contracting over natural resource use in a context of weak property rights: The case of Indonesia. *Environmental and Resource Economics* 33 (1):73-93.

Engel, S., and R. López, 2004. Exploiting Common Resources with Capital-Intensive Technologies: The Role of External Forces. ZEF Discussion Papers on Development Policy 90. Bonn, Germany: Center for Development Research.

Feder, G., and G. Feeny. 1991. Land tenure and property rights: theory and implications for development. *World Bank Economic Review* 5(1): 135-53.

FWI/GFW (Forest Watch Indonesia/Global Forest Watch). 2002. The State of the Forest: Indonesia. Bogor, Indonesia: Forest Watch Indonesia, and Washington DC, US: Global Forest Watch.

Forrester, G., and R.J., May. 1999. *The fall of Suharto.* Singapore: Select Books.

Friend, T. 2003. 'Indonesian destinies', Belknap Press of Harvard University Press, Cambridge, Massachusetts and London.

Gintis, H. 2000. Game theory evolving. *A problem-centered introduction to modeling strategic interaction*. Princeton, US: Princeton University Press.

Greene, W. H. 2000. *Econometric analysis*. New Jersey, US: Prentice Hall.

Hoddinott, J. 2002. Participation and poverty reduction: an analytical framework and overview of the issues. *Journal of African Economies* 2: (1): 146-168.

Hofman, B and Kaiser, K. 2002. The making of the Big Bang and its aftermath: a political economy perspective. Conference paper. Atlanta, US: 'Can decentralization help rebuild Indonesia?', sponsored by the International Studies Program, Andrew Young School of Policy Studies, Georgia State University May 1-3 2002.

Iwan, R. 2004. Mobilizing community conservation. A community initiative to protect its forest against logging in Indonesia. Conference paper. Oaxaca, Mexico: the 10th Biennial Conference of the International Association for the Study of Common Property (IASCP), 9-13 August 2004.

KKPKD (Kelompok Kerja Program Kehutanan Daerah). 2001. *Potret Kehutanan Kabupaten Kutai Barat*. Kelompok Kerja Program Kehutanan Daerah (KKPKD, 'working group programme for forestry') Kabupaten Kutai Barat, Indonesia: Pemerintah Kabupaten Kutai.

Kaimowitz, D., Vallejos, C., Pacheco, P., and R. Lopez. 1998. Municipal governments and forest management in lowland Bolivia. *Journal of Environment and Development* 7 (1): 45-59.

Kompas. 2003. Pengelolaan HPHH di Kutai Barat Gagal Total - Hanya 30 Persen Izin yang Dimanfaatkan (management of HPHH in Kutai Barat – only 30 percent of permits are operational). 4 June 2003. *Kompas* newspaper online version (accessed October 2004): http://www.kompas.com/kompascetak/0306/04/daerah/347777.htm

Larson, A.M. 2004. Democratic decentralization in the forestry sector: lessons learnt from Africa, Asia and Latin America. Conference paper. New York, US: Interlaken Workshop on Decentralization in Forestry (27-30 April 2004), United Nations Forum on Forests (UNFF).

Larson, A.M. and Ribot, J.C. 2004. Democratic decentralization through a natural resource lens: an introduction. *European Journal of Development Research* 16 (1): 1-25.

Limberg, G. 2004. The impact of small concessions on communities in the Malinau watershed. Chapter in preparation for *Decentralization, Forests and Ethnicity*. L.Wollenberg (ed). Bogor, Indonesia: Center for International Forestry Research.

Lindayati, R. 2001. Ideas and institutions in social forestry policy. In *Which way forward? People, forests, and policymaking in Indonesia*. Colfer, C.J.P. and P. Resosudarmo (eds.): Chapter 2. Washington DC, US: Resources for the Future, and Bogor, Indonesia: Center for International Forestry Research and the Institute for Southeast Asian Studies.

McCarthy, J. 2001. Decentralization, local communities and forest management in Bariti Selatan district, Central Kalimantan, Case Study 1. Bogor, Indonesia: Center for International Forestry Research.

McCarthy, J. 2000. Wild Logging': the rise and fall of logging networks and biodiversity conservation projects on Sumatra's rainforest frontier. Occasional Paper no. 31. Bogor, Indonesia: Center for International Forestry Research.

McGrath, and Grandalski. 2000. Forest law enforcement – policies, strategies and technologies. Conference paper. Phnom Penh, Cambodia: Mekong Basin Symposium on Forest Law Enforcement, June 1999.

Mayers, J. and S. Vermeulen. 2002. *Company-community forestry partnerships: From raw deals to mutual gains?* London, UK: Instruments for sustainable private sector forestry series, International Institute for Environment and Development.

Makinuddin, N. 2000. Rattan gardens in conflict with HTI. Workshop paper. Jakarta, Indonesia: PLASMA - Institute for Environment and People Empowerment.

Ministry of Forestry, Republic of Indonesia. 2004. Decentralization of the forestry sector: Indonesia's experience. Conference paper. New York, US: Interlaken Workshop on Decentralization in Forestry (27-30 April 2004), United Nations Forum on Forests (UNFF).

Muthoo, A. 1999. *Bargaining theory with applications.* Cambridge, UK: Cambridge University Press.

National Resources Institute (University of Greenwich, UK). 1996. A review of ODA forestry sector strategy for Indonesia. Jakarta, Indonesia: Overseas Development Administration.

Obidzinski, K. 2004. Illegal logging and the fate of Indonesia's forests in times of regional autonomy. Conference paper. Oaxaca, Mexico: the 10th Biennial Conference of the International Association for the Study of Common Property (IASCP), 9-13 August 2004.

Obidzinski, K. and C. Palmer. 2002. How much do you wanna buy? A methodology for estimating the level of illegal logging in East Kalimantan, Indonesia. Research Report. Bogor, Indonesia: the Centre for International Forestry Research.

Obidzinski, K., Suramenggala, I., and P. Levang. 2001. L'Exploitation Forestiere Illegale en Indonésie: un Inquiétant Processus de Légalisation. *Bois et Forets des Tropiques* 270(4): 85-97.

Ostrom, E. 1990. *Governing the commons: the evolution of institutions for collective action.* Cambridge, UK: Cambridge University Press.

Pagiola, S., and G. Platais. 2002. Payments for Environmental Services. *Environment Strategy Notes* 3 (May 2002). Washington DC, US: World Bank.

Palmer, C. 2004. The role of collective action in determining the benefits from IPPK logging concessions: a case study from Sekatak, East Kalimantan. Working Paper No. 28. Bogor, Indonesia: Center for International Forestry Research.

Palmer, C. 2001. The extent and causes of illegal logging in: an analysis of a major cause of tropical deforestation in Indonesia. Working Paper ISSN 0967-8875. London, UK: The Centre for Social and Economic Research on the Global Environment (CSERGE), University College London.

Palmer, C., and K. Obidzinski. 2002. Indonesia case study prepared for the report: Higher International Standards or Rent-Seeking Race to the Bottom? The Impacts of Forest Trade Liberalisation of Forest Governance. Project GCP/INT/775/JPN. Food and Agriculture Organisation (FAO) and the International Institute for Environment and Development (IIED). Mimeo, Center for Development Research, Bonn, Germany.

Pearce, D. W. 2000. *Environment and Economic Development.* (draft only). Cheltenham, UK: Edward Elgar.

Pearce, D. W., and J., Warford. 1993. *World without end – economics, environment and sustainable development.* New York, US: Oxford University Press.

Peluso, N.L. 1992. The Political Ecology of Extraction and Extractive Reserves in East Kalimantan, Indonesia. *Development and Change* 23 (4): 47-74.

Pemkab Kutai Barat. 2004. Online statistics for the district of Kutai Barat, East Kalimantan province: http://www.kutaibarat.go.id (accessed in April 2004).

Potter, L. 1990. Forest classification, policy and land-use planning in Kalimantan. *Borneo Review* 1(1): 91-128.

Repetto, R, and M. Gillis. 1988. *Public policies and the misuse of forest resources.* Washington DC, US: World Resources Institute, and New York, US: University of Cambridge Press.

Resosudarmo, I.A.P. 2004. Closer to people and trees: will decentralisation work for the people and the forests of Indonesia? *European Journal of Development Research* 16 (1): 110-32.

Rhee, S. 2000. De facto decentralization and the management of natural resources in East Kalimantan during a period of transition. *Asia Pacific Community Forestry Newsletter* 13 (2): 34-9.

Ribot, J.C. 2004. Choosing representation: institutions and powers for decentralized natural resource management. Conference paper. New York, US: Interlaken Workshop on Decentralization in Forestry (27-30 April 2004), United Nations Forum on Forests (UNFF).

Ribot, J.C. 2002. African decentralization: local actors, powers and accountability. United Nations Research Institute on Social Development (UNRISD) Programme on Democracy, Governance and Human Rights, Paper No. 8. Geneva, Switzerland: UNRISD.

Richards, M., C. Palmer, C.F. Young, and K. Obidzinski. 2003. Higher International Standards or Rent-Seeking Race to the Bottom? The Impacts of Forest Trade Liberalisation of Forest Governance. Project GCP/INT/775/JPN. Rome, Italy: Food and Agriculture Organisation (FAO).

Ross, M. 2001. *Timber Booms and Institutional Breakdown in Southeast Asia.* Cambridge, UK: Cambridge University Press.

Ruzicka, I. 1979. Economic aspects of Indonesian timber concessions 1967-76. University of London, UK: Ph.D. dissertation, mimeo.

Schwarz, A. 1990. A saw point for ecology. *Far Eastern Economic Review* 148 (16): 60.

Scotland, N., A. Fraser., and N. Jewell. 1999. Roundwood supply and demand in the forest sector in Indonesia. Report No. PFM/EC. 99/08. Jakarta, Indonesia: Indonesia-UK Tropical Forest Management Programme.

Smith J. and K. Obidzinski. 2002. Moving toward strategies to combat illegal logging. Conference paper. Yale, US: conference on 'Illegal logging in the tropics – politics and economy of resource misuse', Yale School of Forestry, Yale University, March 29-31 2002.

Southeast Asia Sustainable Forest Management Network (SASFMN). 1993. Communities and Forest Management in East Kalimantan: Pathway to Environmental Stability. Research Network Report Number 3, August 1993. Berkeley, US: Centre for Southeast Asia Studies, University of California, Berkeley.

Sudana, M. 2004. Winner takes all. Understanding forest conflict in the era of decentralization in Indonesia. Conference paper. Oaxaca, Mexico: the 10th Biennial Conference of the International Association for the Study of Common Property (IASCP), 9-13 August 2004.

Suramenggala, I., Sutisna, I.H., Saragih, I.B., Samsu, Ngau, Y. 2001. Evaluasi dampak IPPK terdahap fisik hutan, pendapaten masyarakat dan permerintah kabupatan Bulungan (An evaluation of the impacts of IPPK on forests, community incomes and the government of Bulungan district). East Kalimantan, Indonesia: Yayasan Pionir Bulungan.

UNDP. 2003. United Nations Development Programme Human Development Report for Indonesia. New York, US: United Nations.

Van Zorge Report. 1999. Decentralizing amid regional dissent. Jakarta, Indonesia: 26 Nov, pp. 4-13.

Wollenberg, E. 2004. Riding the rapids: synthesis and conclusion. Draft chapter in preparation for *Decentralization, Forests and Ethnicity.* Wollenberg, L. (ed): Chapter 12. Bogor, Indonesia: Center for International Forestry Research.

Wollenberg, E., and H. Kartodihardjo. 2002. Devolution and Indonesia's new forestry law. In *Which way forward? People, forests and policymaking in*

Indonesia. Colfer, C.J.P. and P. Resosudarmo (eds.): 81-109. Washington DC, US: Resources for the Future, and Bogor, Indonesia: Center for International Forestry Research and Institute for Southeast Asian Studies.

World Bank. 2000. *The challenges of World Bank involvement in forests: an evaluation of Indonesia's forests and World Bank assistance*. Washington DC, US: World Bank.

World Resources Institute (WRI). 2000. *Trial by fire: forest fires in Indonesia's era of crisis and reform*. Washington DC, US: World Resources Institute.

Appendices

Appendix 1 Paired Samples Test for Differences in Proportions of Households receiving Financial and Non-monetary Benefits[1], from Logging Companies before and after Decentralization

Pairs	Proportion of households receiving benefits	Paired differences					T	Df	Sig. (2-tailed)
		Mean (%)	*Std. Deviation*	*Std. Error Mean*	95% confidence interval of the difference				
					Lower	*Upper*			
1	Money	-92.72	17.68	2.28	-97.28	-88.15	-40.621	59	**0.000
2	Non-monetary developments	-6.85	5.61	5.89	-8.14	-5.56	-3.211	59	**0.000

Note 1 – includes perception of developments provided at the community level in addition to those received individually by households. Money received at community level is supposedly either given to individual households or spent on public goods. *difference is significant at 0.10 level; ** difference is significant at 0.05 level.

Appendix 2 Household Perceptions of Forest Ownership before and after Decentralization and the Results from Paired Samples Test for Differences in Perceptions

Who owned/owns the forest?	Before 1997-98		2003-04		Mean change in response[1]
	Number	*% of house-holds*	*Number*	*% of house-holds*	
Community	126	20.9	494	81.9	+ 61.0**
Community-state	18	3.0	27	4.5	+ 1.5*
Community-firm	1	0.2	8	1.3	+ 1.2*
Community-other	1	0.2	10	1.7	+ 1.5*
State	233	38.6	18	3.0	- 35.7**
State-firm	121	20.1	0	0	- 20.1**
State-other	0	0	0	0	0
Firm	67	11.1	2	0.3	- 10.8**
Firm-other	0	0	0	0	0
Other	6	1.0	11	1.8	+ 0.8
Community-state-firm	3	0.5	23	3.8	+ 3.3*
Community-state-other	0	0	3	0.5	+ 0.5*
Community-state-firm-other	0	0	1	0.2	+ 0.2
Don't know	35	5.8	45	7.5	+ 1.7

Note In the questionnaire, there was no distinction made between central and local government before 1997-98 mainly because local governments hardly existed at that time. Responses from 603 households were included. 1. Paired samples test was used to test for differences in households perceptions of forest ownership before and after decentralization, the detailed results for which can be seen in appendix. *difference is significant at 0.10 level; ** difference is significant at 0.05 level.

Appendix 3 Correlation Matrix for Factor Analysis using Discrete Variable for Firm Compliance

		FIRMNONC	*INTRACOM*	*INCCONFF*	*MAXCONFC*	*HHPARCON*
Correlation	*FIRMNONC*	1.000	0.440	0.539	0.313	0.498
	INTRACOM	0.440	1.000	0.216	0.075	0.078
	INCCONFF	0.539	0.216	1.000	0.499	0.682
	MAXCONFC	0.313	0.075	0.499	1.000	0.588
	HHPARCON	0.498	0.078	0.682	0.588	1.000
Sig. (1-tailed)	*FIRMNONC*		0.000	0.000	0.007	0.000
	INTRACOM	0.000		0.049	0.285	0.276
	INCCONFF	0.000	0.049		0.000	0.000
	MAXCONFC	0.007	0.285	0.000		0.000
	HHPARCON	0.000	0.276	0.000	0.000	

Note Determinant of correlation matrix = 0.178

Appendix 4 Correlation Matrix for Factor Analysis using Continuous Variable for Firm Compliance

		INTRACOM	*INCCONFF*	*MAXCONFC*	*HHPARCON*	*FEECOMP2*
Correlation	*INTRACOM*	1.000	0.297	0.060	0.093	- 0.104
	INCCONFF	0.297	1.000	0.499	0.682	- 0.212
	MAXCONFC	0.060	0.499	1.000	0.588	0.136
	HHPARCON	0.093	0.682	0.588	1.000	0.028
	FEECOMP2	- 0.104	- 0.212	0.136	0.028	1.000
Sig. (1-tailed)	*INTRACOM*		0.011	0.325	0.241	0.214
	INCCONFF	0.011		0.000	0.000	0.052
	MAXCONFC	0.325	0.000		0.000	0.150
	HHPARCON	0.241	0.000	0.000		0.416
	FEECOMP2	0.214	0.052	0.150	0.416	

Note Determinant of correlation matrix = 0.259

Appendix 5 Paired Samples Test for Differences Local Environmental Impacts from Mechanised Logging before and after Decentralization.

Pairs	Logging impact	Paired differences					t	df	Sig. (2-tailed)
		Mean[1]	*Std. Deviation*	*Std. Error Mean*	*95% confidence interval of the difference*				
					Lower	*Upper*			
1	Water quality	-0.642	26.690	3.446	-7.536	6.253	-0.186	59	0.853
2	Flooding	5.600	27.870	3.598	-1.600	12.800	1.556	59	0.125
3	Hunting	1.165	24.196	3.124	-5.085	7.415	0.373	59	0.711
4	Forest products	23.153	35.693	4.647	13.851	32.454	4.982	58	**0.000
5	Farming	13.312	21.905	2.828	7.653	18.970	4.707	59	**0.000

Note *significant at 0.10 level; **significant at 0.05 level. 1. The average difference in (negative) responses given for impacts before and after decentralization for all communities.

Appendix 6 Descriptive Statistics on Parameter Proxies

Variable	Mean	Std.Dev.	Min	Max
Value of payment plus social provision	32789	24131	2500	106322
No. of years of community residence in current location	78.0	61.8	12.0	300.0
Dominant ethnic group is Dayak Punan	0.1	0.27	0	1.0
No. of neighbouring communities sharing forest border	2.7	1.2	1.0	8.0
Forest claim per capita (Ha)	301.5	528	1.1	2778
Length of exposure to commercial logging prior to negotiations (years)	19.1	9.7	0	33.0
Average % of household incomes derived from sale of forest products	39.7	25.1	5.0	83.3
Forest quality: area logged before commercial operation	0.5	0.5	0	1.0
Only community leaders knew about negotiation system before negotiations commenced	0.6	0.5	0	1.0
Actual area logged (ha)	1827	2127	200.0	15600
NGO Assistance	0.2	0.4	0	1.0
Community located in Kutai Barat district	0.3	0.5	0	1.0
Number of households	166.3	151.6	17.0	950
All groups in community can participate in community organizations	0.6	0.5	0	1.0
% of households that participate in community organizations	30.2	18.6	0	78.0
% of households containing members of dominant ethnic grouping	82.6	15.9	40.0	99.0
Number of ethnic groups in community with 20 percent or more of total population	1.3	0.5	1.00	3.0
Distance to nearest market (km)	22.9	31.4	0	150.0
% of households containing at least one government employee	6.1	7.4	0	41.3
Electricity	0.6	0.5	0	1.0
Health clinic	0.6	0.5	0	1.0
School	0.9	0.3	0	1.0
Rice mill	0.7	0.5	0	1.0
% of hhs with televisions before agreement	16.6	16.7	0	60.0
% of hhs owning mode of transport before agreement	48.9	24.0	0	100.0
% of hhs holding savings before agreement	15.9	18.7	0	60.0
% of hhs with schooling after compulsory period (aged 15)	39.3	24.5	0	100.0
Previous experience negotiating with loggers prior to current negotiations	0.1	0.4	0	1.0
Broker known to community	0.5	0.5	0	1.0
% of hhs with previous experience of working for logging firms	49.0	22.1	0	83.0

Appendix 7 *Hypotheses 1 & 2.* Proxies Used, Expected Effects on Community Payoffs, and Two-sided Correlations between Community Payoffs and Explanatory Variables

Parameter/ Proxy	Definition	Probability of community able to enforce property rights		Payment to community (for those able to enforce property rights)	
		Exp. effect on	2-sided correlation with comm. payoffs	Exp. effect on	2-sided correlation with comm. payoffs
B	***Community valuation of standing forest***	+			
YRLOCATE	No. of years of community residence in current location	+	0.079	+	-0.080
PUNAN	Dominant ethnic group is Dayak Punan (semi-nomadic ethnic group) (Yes=1, No=0)	-	-0.189	-	-0.226*
NONECOMM	No. of neighbouring communities sharing forest border	-	-0.250**	-	0.051
FORCLPH	Forest claim per capita (Ha)	-	-0.041	-	-0.022
YRSHPH	Length of exposure to commercial logging activities prior to negotiations (years)	+	0.065	+	0.198*
NGOASS	NGO assistance: advice, information with respect to logging impacts and agreements (Yes=1,No=0)	+	-0.087	+	-0.196
AVINCFP	Average % of household incomes derived from sale of forest products	+	0.316**	+	0.300**
ELITEKNW	Only community leaders knew about negotiation system before negotiations commenced (Yes=1, No=0)	NA		-	0.001
V	***Profitability of logging***	-		+	
ACTHA	Actual area logged (ha)	?	-0.301**	?	-0.172
BROKOUT	Use of a broker from outside the community during negotiations (Yes=1, No=0)	+	-0.159	-	-0.346**
KUTAIBAR	Community located in Kutai Barat district (Yes=1, No=0)	+	0.309**	-	0.510**
LOGGEDB4	Forest quality: area logged before by commercial operation (Yes=1, No=0)	+	0.315**	-	0.247*

S	***Community blockading costs***	-		*NA*	
NOHH	Number of households	-	0.380	NA	
PARTORG	All groups in community can participate in community organizations (Yes=1, No=0))	+	0.141	NA	
HPARTORG	% of households that participate in community organizations	+	0.156	NA	
PHHDOMGP	% of households containing members of dominant ethnic grouping	+	0.178*	NA	
NOGP20	Number of ethnic groups in community with 20 percent or more of total population	-	-0.167	NA	
DISTMARK	Distance to nearest market (km)	?	-0.138	NA	
PGOVJOB	% of households containing at least one government employee	-	-0.069	NA	
r^C	***Community discount rate***	-		-	
ELECTRIC	Electricity (Yes=1, No=0)	+	0.002	+	-0.035
HEALTH	Health clinic (Yes=1, No=0)	+	0.073	+	0.075
SCHOOL	School (Yes=1, No=0)	+	0.162	+	0.080
RICEMILL	Rice mill (Yes=1, No=0)	+	0.215*	+	0.146
HOWNTV	Prop. of hhs with televisions before agreement	+	0.265**	+	0.072
HOWNTRAN	% of hhs owning mode of transport before agreement	+	0.145	+	-0.005
HBANKACC	% of hhs holding savings before agreement	+	0.210*	+	0.124
T	***Community's bargaining power***	**NA**		+	
NEGHPH	Previous experience negotiating with loggers prior to current negotiations (Yes=1, No=0)	NA		+	0.154
BROKKNW	Broker known to community (Yes=1, No=0)	NA		+	0.127
EXPLOGCO	% of hhs with previous experience of working for logging companies	NA		+	0.234**

Note *significant at 0.10 level; **significant at 0.05 level. NA denotes not applicable.

Appendix 8 *Hypothesis 3.* Proxies Used, Expected Effects on Community Payoffs, and Two-sided Correlations between Community Payoffs and Explanatory Variables

Parameter/ Proxy	Definition	Expected effect on value of payment & in-kind provision to community	Two-sided correlation with community payoffs	
			Fee (ACTFEE)	*Fee + non-monetary (ACTFEE DE)*
B	***Community valuation of standing forest***	+		
YRLOCATE	No. of years of community residence in current location	+	0.021	0.013
PUNAN	Dominant ethnic group is Dayak Punan (semi-nomadic ethnic group) (Yes=1, No=0)	-	-0.159	-0.157
NONECOMM	No. of neighbouring communities sharing forest border	-	-0.136	-0.125
FORCLPH	Forest claim per capita (Ha)	-	-0.055	-0.051
YRSHPH	Length of exposure to commercial logging activities prior to negotiations (years)	+	0.127	0.129
AVINCFP	Average % of household incomes derived from sale of forest products	+	0.339**	0.348**
ELITEKNW	Only community leaders knew about negotiation system before negotiations commenced (Yes=1, No=0)	-	-0.065	-0.055
V	***Profitability of logging***	?		
ACTHA	Actual area logged (ha)	?	0.283**	0.282**
KUTAIBAR	Community located in Kutai Barat district (Yes=1, No=0)	?	0.443**	0.446**
LOGGEDB4	Forest quality: area logged before by commercial operation (Yes=1, No=0)	?	0.305**	0.312**
S	***Community blockading costs***	+		
NOHH	Number of households	-	0.014	0.021
PARTORG	All groups in community can participate in community organizations (Yes=1, No=0))	+	0.159	0.149
HPARTORG	% of households that participate in community organizations	+	0.018	0.016

S	***Community blockading costs (cont'd)***	+		
PHHDOMGP	% of households containing members of dominant ethnic grouping	+	0.177*	0.176*
NOGP20	Number of ethnic groups in community with 20 percent or more of total population	-	-0.206**	-0.209**
DISTMARK	Distance to nearest market (km)	?	-0.211**	-0.200**
PGOVJOB	% of households containing at least one government employee	-	-0.047	-0.037
r^C	***Community discount rate***	-		
ELECTRIC	Electricity (Yes=1, No=0)	+	-0.043	-0.028
HEALTH	Health clinic (Yes=1, No=0)	+	0.078	0.085
SCHOOL	School (Yes=1, No=0)	+	0.113	0.105
RICEMILL	Rice mill (Yes=1, No=0)	+	0.177*	0.175*
HSCHO15	% of households with schooling after compulsory period (aged 15)	+	0.097	0.100
HOWNTV	% of hhs with televisions before agreement	+	0.214**	0.225**
HOWNTRAN	% Of hhs owning mode of transport before agreement	+	0.096	0.108
HBANKACC	% of hhs holding savings before agreement	+	0.202**	0.205**
T	***Community's bargaining power***	+		
NEGHPH	Previous experience negotiating with loggers prior to current negotiations (Yes=1, No=0)	+	0.139	0.141
BROKKNW	Broker known to community (Yes=1, No=0)	+	-0.109	-0.101
EXPLOGCO	% of hhs with previous experience of working for logging companies	+	0.094	0.096

Note *significant at 0.10 level; **significant at 0.05 level.

Development Economics and Policy

Series edited by Franz Heidhues and Joachim von Braun

Band 1 Andrea Fadani: Agricultural Price Policy and Export and Food Production in Cameroon. A Farming Systems Analysis of Pricing Policies. The Case of Coffee-Based Farming Systems. 1999.

Band 2 Heike Michelsen: Auswirkungen der Währungsunion auf den Strukturanpassungsprozeß der Länder der afrikanischen Franc-Zone. 1995.

Band 3 Stephan Bea: Direktinvestitionen in Entwicklungsländern. Auswirkungen von Stabilisierungsmaßnahmen und Strukturreformen in Mexiko. 1995.

Band 4 Franz Heidhues / François Kamajou: Agricultural Policy Analysis – Proceedings of an International Seminar, held at the University of Dschang, Cameroon on May 26 and 27 1994, funded by the European Union under the Science and Technology Program (STD). 1996.

Band 5 Elke M. Förster: Protection or Liberalization? A Policy Analysis of the Korean Beef Sector. 1996.

Band 6 Gertrud Schrieder: The Role of Rural Finance for Food Security of the Poor in Cameroon. 1996.

Band 7 Nestor R. Ahoyo Adjovi: Economie des Systèmes de Production intégrant la Culture de Riz au Sud du Bénin: Potentialités, Contraintes et Perspectives. 1996.

Band 8 Jenny Müller: Income Distribution in the Agricultural Sector of Thailand. Empirical Analysis and Policy Options. 1996.

Band 9 Michael Brüntrup: Agricultural Price Policy and its Impact on Production, Income, Employment and the Adoption of Innovations. A Farming Systems Based Analysis of Cotton Policy in Northern Benin. 1997.

Band 10 Justin Bomda: Déterminants de l'Epargne et du Crédit, et leurs Implications pour le Développement du Système Financier Rural au Cameroun. 1998.

Band 11 John M. Msuya: Nutrition Improvement Projects in Tanzania: Implementation, Determinants of Performance, and Policy Implications. 1998.

Band 12 Andreas Neef: Auswirkungen von Bodenrechtswandel auf Ressourcennutzung und wirtschaftliches Verhalten von Kleinbauern in Niger und Benin. 1999.

Band 13 Susanna Wolf (ed.): The Future of EU-ACP Relations. 1999.

Band 14 Franz Heidhues / Gertrud Schrieder (eds.): Romania – Rural Finance in Transition Economies. 2000.

Band 15 Katinka Weinberger: Women's Participation. An Economic Analysis in Rural Chad and Pakistan. 2000.

Band 16 Christof Batzlen: Migration and Economic Development. Remittances and Investments in South Asia: A Case Study of Pakistan. 2000.

Band 17 Matin Qaim: Potential Impacts of Crop Biotechnology in Developing Countries. 2000.

Band 18 Jean Senahoun: Programmes d'ajustement structurel, sécurité alimentaire et durabilité agricole. Une approche d'analyse intégrée, appliquée au Bénin. 2001.

Band 19 Torsten Feldbrügge: Economics of Emergency Relief Management in Developing Countries. With Case Studies on Food Relief in Angola and Mozambique. 2001.

Band 20 Claudia Ringler: Optimal Allocation and Use of Water Resources in the Mekong River Basin: Multi-Country and Intersectoral Analyses. 2001.

Band 21 Arnim Kuhn: Handelskosten und regionale (Des-)Integration. Russlands Agrarmärkte in der Transformation. 2001.

Band 22 Ortrun Anne Gronski: Stock Markets and Economic Growth. Evidence from South Africa. 2001.

Band 23 Patrick Webb / Katinka Weinberger (eds.): Women Farmers. Enhancing Rights, Recognition and Productivity. 2001.

Band 24 Mingzhi Sheng: Lebensmittelkonsum und -konsumtrends in China. Eine empirische Analyse auf der Basis ökonometrischer Nachfragemodelle. 2002.

Band 25 Maria Iskandarani: Economics of Household Water Security in Jordan. 2002.

Band 26 Romeo Bertolini: Telecommunication Services in Sub-Saharan Africa. An Analysis of Access and Use in the Southern Volta Region in Ghana. 2002.

Band 27 Dietrich Müller-Falcke: Use and Impact of Information and Communication Technologies in Developing Countries' Small Businesses. Evidence from Indian Small Scale Industry. 2002.

Band 28 Wolfram Erhardt: Financial Markets for Small Enterprises in Urban and Rural Northern Thailand. Empirical Analysis on the Demand for and Supply of Financial Services, with Particular Emphasis on the Determinants of Credit Access and Borrower Transaction Costs. 2002.

Band 29 Wensheng Wang: The Impact of Information and Communication Technologies on Farm Households in China. 2002.

Band 30 Shyamal K. Chowdhury: Institutional and Welfare Aspects of the Provision and Use of Information and Communication Technologies in the Rural Areas of Bangladesh and Peru. 2002.

Band 31 Annette Luibrand: Transition in Vietnam. Impact of the Rural Reform Process on an Ethnic Minority. 2002.

Band 32 Felix Ankomah Asante: Economic Analysis of Decentralisation in Rural Ghana. 2003.

Band 33 Chodechai Suwanaporn: Determinants of Bank Lending in Thailand: An Empirical Examination for the Years 1992 to 1996. 2003.

Band 34 Abay Asfaw: Costs of Illness, Demand for Medical Care, and the Prospect of Community Health Insurance Schemes in the Rural Areas of Ethiopia. 2003.

Band 35 Gi-Soon Song: The Impact of Information and Communication Technologies (ICTs) on Rural Households. A Holistic Approach Applied to the Case of Lao People's Democratic Republic. 2003.

Band 36 Daniela Lohlein: An Economic Analysis of Public Good Provision in Rural Russia. The Case of Education and Health Care. 2003.

Band 37 Johannes Woelcke. Bio-Economics of Sustainable Land Management in Uganda. 2003.

Band 38 Susanne M. Ziemek: The Economics of Volunteer Labor Supply. An Application to Countries of a Different Development Level. 2003.

Band 39 Doris Wiesmann: An International Nutrition Index. Concept and Analyses of Food Insecurity and Undernutrition at Country Levels. 2004.

Band 40 Isaac Osei-Akoto: The Economics of Rural Health Insurance. The Effects of Formal and Informal Risk-Sharing Schemes in Ghana. 2004.

Band 41 Yuansheng Jiang: Health Insurance Demand and Health Risk Management in Rural China. 2004.

Band 42 Roukayatou Zimmermann: Biotechnology and Value-added Traits in Food Crops: Relevance for Developing Countries and Economic Analyses. 2004.

Band 43 F. Markus Kaiser: Incentives in Community-based Health Insurance Schemes. 2004.

Band 44 Thomas Herzfeld: *Corruption begets Corruption*. Zur Dynamik und Persistenz der Korruption. 2004.

Band 45 Edilegnaw Wale Zegeye: The Economics of On-Farm Conservation of Crop Diversity in Ethiopia: Incentives, Attribute Preferences and Opportunity Costs of Maintaining Local Varieties of Crops. 2004.

Band 46 Adama Konseiga: Regional Integration Beyond the Traditional Trade Benefits: Labor Mobility contribution. The Case of Burkina Faso and Côte d'Ivoire. 2005.

Band 47 Beyene Tadesse Ferenji: The Impact of Policy Reform and Institutional Transformation on Agricultural Performance. An Economic Study of Ethiopian Agriculture. 2005.

Band 48 Sabine Daude: Agricultural Trade Liberalization in the WTO and Its Poverty Implications. A Study of Rural Households in Northern Vietnam. 2005.

Band 49 Kadir Osman Gyasi: Determinants of Success of Collective Action on Local Commons. An Empirical Analysis of Community-Based Irrigation Management in Northern Ghana. 2005.

Band 50 Borbala E. Balint: Determinants of Commercial Orientation and Sustainability of Agricultural Production of the Individual Farms in Romania. 2006.

Band 51 Pamela Marinda: Effects of Gender Inequality in Resource Ownership and Access on Household Welfare and Food Security in Kenya. A Case Study of West Pokot District. 2006.

Band 52 Charles Palmer: The Outcomes and their Determinants from Community-Company Contracting over Forest Use in Post-Decentralization Indonesia. 2006.

www.peterlang.de

FSC
www.fsc.org
MIX
Papier | Fördert
gute Waldnutzung
FSC® C083411

Kolibri 360 GmbH
Ferdinand-Jühlke-Straße 7
99095 Erfurt, Deutschland
produktsicherheit@kolibri360.de

Druck:
CPI Druckdienstleistungen GmbH
im Auftrag der
Kolibri 360 GmbH
Ferdinand-Jühlke-Str. 7
99095 Erfurt

The Outcomes and their Determinants from Community-Company Contracting over Forest Use in Post-Decentralization Indonesia

DEVELOPMENT ECONOMICS AND POLICY

Series edited by Franz Heidhues and Joachim von Braun

Vol. 52

PETER LANG

Frankfurt am Main · Berlin · Bern · Bruxelles · New York · Oxford · Wien